SHEEP

SHEEP

Small-Scale Sheep Keeping

by Sue Weaver

PRESS

Sheep

Project Team
Assistant Production Manager: Tracy Vogtman
Editors: Jennifer Calvert, Amy Deputato,
Lindsay Hanks, Karen Julian, Jarelle S. Stein
Cover Design: Cindy Kassebaum,
Book Design Concept: Lisa Barfield
Book Design and Layout: Allyn A. Salmond

I-5 PUBLISHING, LLC™
Chief Executive Officer: Mark Harris
Chief Financial Officer: Nicole Fabian
Vice President, Chief Content Officer: June Kikuchi
General Manager, I-5 Press: Christopher Reggio
Editorial Director, I-5 Press: Andrew DePrisco
Art Director, I-5 Press: Mary Ann Kahn
Digital General Manager: Melissa Kauffman
Production Director: Laurie Panaggio
Production Manager: Jessica Jaensch
Marketing Director: Lisa MacDonald

The Library of Congress has cataloged an earlier edition as follows:
Weaver, Sue.
Sheep: small-scale sheep keeping for pleasure and profit / Sue Weaver.
p. cm.——(Hobby farms)
Includes bibliographical references.
ISBN 1-931993-49-1
1. Sheep. 2. Sheep—United States. I. Title. II. Series.
SF375.W368 2005
636.3—dc22
2004025520

This book has been published with the intent to provide accurate and authoritative information in regard to the subject matter within. While every precaution has been taken in the preparation of this book, the author and publisher expressly disclaim any responsibility for any errors, omissions, or adverse effects arising from the use or application of the information contained herein. The techniques and suggestions are used at the reader's discretion and are not to be considered a substitute for veterinary care. If you suspect a medical problem, consult your veterinarian.

I-5 Publishing, LLC™
3 Burroughs, Irvine, CA 92618
www.facebook.com/i5press
www.i5publishing.com

Printed and bound in China
13 14 15 16   5 7 9 8 6

This book is dedicated to Baasha (Brighton Ridge Farms #A-80), the matriarch of our flock and a really neat sheep…

…to Dick Harward and Barb Zebbs of Wind Fall Farm in Willow Springs, Missouri, for allowing us to photograph their beautiful Scottish Blackface sheep.

…and to Laurie, Cathy, Lyn, Lynn, Barbara, Marion, Kim, Dawn, Bernadette, Kris, Melissa, Connie, Kelly, Kathy, Liz, Lisa, Michelle, and all the other shepherds at the Hobby Farms Sheep E-mail discussion group. Thanks for your help, ladies!

Contents

Why Sheep?

Sheep are to hobby farms as diamonds are to gold: they make a good thing better. Be they pets or profit makers, sheep should be part of every small-farm scene. They are inexpensive to buy and keep, easy to care for, and relatively long lived, making them great investments. Given predator-proof fencing, minimal housing, good feed, and a modicum of daily attention, sheep will thrive. New shepherds can learn to care for sheep in a relatively short period of time, which makes them attractive even to first-time farmers. Sheep make delightful pets; they'll mow your yard, come when called, and with training, maybe even pull a cart.

Farmers looking to turn a profit in sheep can do so. Hand-tamed miniature lambs fetch respectable prices as pets. Clean fleeces from many breeds are wildly popular among handspinners. Niche-marketed lamb sells readily to ethnic and organic markets, and sheep's milk cheese is popular with gourmets around the world. In addition, raising livestock qualifies landowners for lower agricultural land tax assessments in many locales.

You may have reservations; even I haven't had sheep all my life. For more than fifty years I was blissfully content with husband, horses, and dogs. Then came hobby farms—then guineas, chickens, and an ox. I had always liked sheep, so why not give them a try? I was smitten, totally and thoroughly captivated by our first woolly pets. Now sheep are a part of me. I've been entrusted with the care of these beautiful creatures. I'm sure I've been blessed!

CHAPTER ONE

Sheep from the Beginning

In the beginning, there were majestic wild sheep called mouflons. Hunters stalked the wily sheep to dine on their tasty flesh and to craft cozy clothing from their hides. About 11,000 years ago—probably near Zawi Chemi Shanidar, in what is now northern Iraq—a hunter stopped fiddling with his spear point, kicked a log onto the fire, and said to a friend, "Wouldn't it be smarter to snatch some lambs and raise them here by camp?"

So humans and sheep formed an alliance. People protected sheep from wolves, bears, and mountain lions; sheep reciprocated by developing wool. About 3500 BC, women puzzled out how to weave sheep's woolly covering into fine, sturdy cloth that kept wearers toasty in the wintertime and cool under the blazing summer sun. Decked out in woolen garments, men said to one another, "We don't have to stay here on the Mesopotamian plains where it's always pretty warm; we could go out and conquer the world!"

Sheep were already out in the world. Domestic sheep had reached parts of Europe by 5000 BC, having been carried west by intrepid Neolithic farmers. (Sheep remains have been recovered from a Swiss New Stone Age dig circa 2000 BC.) Swedish farmers began raising northern short-tailed sheep between 4000 and 3000 BC. Between 1000 BC and AD 1, Persians, Greeks, and Romans labored to develop new and better sheep. The Romans brought their revamped woollies along (a walking food supply) when they conquered Europe and North Africa; by AD 50, the Romans had erected a wool-processing plant near Winchester, England.

Historically, the greatest sheep was the mighty Merino. Some researchers think it sprang from a genetic mutation some 3,000 years ago; others believe it was developed during the reign of Queen Claudia of Spain (AD 41–54).

Whatever the case, income from the Spanish Merino wool trade transformed Spain into a world power and financed its New World voyages. Until the mid-eighteenth century, in fact, Spain so hoarded Merinos that it made smuggling sheep out of the country punishable by death.

When Columbus embarked on his second voyage, in 1493, he packed along big, meaty Spanish Churra sheep. He left some in Cuba and more in Santo Domingo. Their descendants trailed Cortez and his conquistadors as they pillaged their way across the New World.

This vintage European Easter card is filled with historical symbols of the season: children in traditional dress, pussy willow boughs, springtime flowers, and fluffy sheep.

Meanwhile, another wave of sheep arrived by way of the North American colonies. Fifteen years after settling Plymouth Colony, the Pilgrims purchased sheep from Dutch dealers on Manhattan Island. By 1643, there were 1,000 sheep in Massachusetts Bay Colony alone. Records show that Governor Winthrop, of the Connecticut colony, acquired a handsome flock of Southdown sheep in 1646. By 1664, an estimated 100,000 sheep called the thirteen colonies home.

Trafficking in sheep or wool was risky business. By 1698, Americans were peddling their wool abroad—much to the consternation of the British king William III. William ultimately outlawed the production of sheep and wool in the colonies. Miscreants caught engaging in the trade had their right hands amputated.

Yet intrepid shepherds continued raising sheep before, during, and after the American Revolution. George Washington and Thomas Jefferson were inaugurated in suits crafted of pure American wool. Both presidents were, in fact, avid sheepmen. Washington raised English Leicesters at Mount Vernon; Jefferson bred English Leicester and Tunis sheep at Monticello.

During the nineteenth century, a slew of European breeds appeared on the American scene. Coveted Spanish Merinos (1808), Lincolns (1825), Cotswolds (1832), Shropshires (1855), and Hampshires (1885) arrived and flourished. In 1912, the first all-American breed, the Columbia, was developed.

SHEEP IN MYTHS

Back to the Dawn of Time:

It stands to reason, considering humanity's long association with sheep, that myth and religion embrace them, too. The Egyptian sun god, Amon-Ra, was depicted as either a ram-headed deity or a sun disc with ram's horns. Other ram-horned deities include the Middle Eastern great goddess Ishtar; the Phoenician sun god Baal-Hamon; and Ea-Oannes, the Babylonian god of the deeps.

The Greek goddess of crossroads, Hecate, is associated with sheep—especially black ewe lambs. Pan was the god of sheep and flocks. A famous mythical sheep was the golden ram, Khrysomallos, a wooly son of Neptune and Theophane, conceived when they appeared in the forms of a ram and a ewe. Flying Khrysomallos carried the children Phrixos and Helle to Kolkhis, where Phrixos sacrificed Khrysomallos to the gods and hung his fleece in the holy grove of Ares. It became the object of Jason and the Argonauts' quest for the golden fleece.

The Romans had Palas, the guardian of their flocks. On his feast day (the Parilia, April 21) sheepfolds were decked with greenery, and a wreath was placed on every entrance. Chuku, supreme deity of the Ibo in Nigeria, once sent a messenger sheep to tell humans that the dead should be placed on the earth and have ashes sprinkled over them; then, they would come back to life. But the sheep forgot the message and decided to wing it, directing instead that the dead should be buried in the ground.

Fairy lore is rife with sheepy connections. The Scots-Irish shape-shifting buachailleen play pranks on shepherds, such as spooking the sheep or smearing their fleeces with muck the night before shearing. Iron bells suspended from collars around sheep's necks protect them from the buachailleen. According to Welsh fairy folk, sheep are the only creatures allowed to graze the grass growing in fairy rings. That, according to legend, is what makes Welsh mutton the best in the world.

America's sheep population peaked in 1942, at a mind-boggling 56.2 million head. Today there are 6.35 million head, and that's down a hefty 14 percent since 2001. The good news is that although large-scale commercial sheep operations are faltering, there is a burgeoning unmet market for specialty products such as handspinners' fleece, gourmet sheep's milk cheeses, and certified organic lamb—products raised on, and marketed from, today's small farms.

Sheep come in many sizes, shapes, colors, and temperaments. They can be classified by type of fleece produced, appearance, or place of origin. Consequently, they offer a wide variety of characteristics to breeders and consumers alike. However, supplies are not unlimited, and some breeds are considered endangered.

SHEEP AT A GLANCE

Domestic sheep (*Ovis aries*) belong to the Bovidae family, along with other hollow-horned, cloven-hoofed ruminants such as cattle, and to the Caprinae subfamily, in the company of their cousins, the goats. There are more than 1,000 breeds of domesticated sheep in the world today—more than three score of them in North America alone. While size, shape, type of fleece (or lack thereof), and disposition vary greatly, all domestic sheep have certain traits in common. We'll discuss many of these items later. For now, here are sheep at a glance.

SHEEP IQ

The University of Illinois monograph "An Introduction to Sheep Behavior" ranks sheep IQ a smidge below that of the pig and on a level with cattle.

Packaged as tightly as canned sardines, these white-faced sheep are comforted by their close quarters. Sheep have a natural flocking tendency and stick together as means of protection.

Researchers at the Babraham Institute in Cambridge, England, trained twenty sheep to recognize pictures of other sheep faces. Electrodes measuring their brain activity proved that some remembered at least fifty of the faces for up to two years. "It's a very sophisticated memory system," explains Dr. Keith Kendrick. "They are showing similar abilities in many ways to humans."

Sheep also learn and respond to their names. Club lambs and exhibition sheep lead, stand tied, allow extensive grooming, and pose in the show ring. Pet sheep learn to pull carts; some even do tricks. Sheep, intelligently and quietly handled, are very trainable.

Sheep are not stupid; they are reactive. Their only means of survival is to band together for protection, then to run. Frightened, stressed sheep flee blindly, pack into corners, and get wedged behind gates. Quietly handled sheep generally do not.

FLOCKING INSTINCT AND SOCIAL STRUCTURE

Sheep are gregarious, meaning they crowd together for reassurance and protection. They have a strong inner compulsion to follow a leader. These traits compose their flocking instinct. In most cases, the leader is simply the first sheep that starts moving in a given direction; flock hierarchy rarely enters the picture.

White-faced (wool) sheep are more gregarious than are black-faced (meat) breeds. When stressed, huge flocks of Australian Merinos can pack so tightly that humans swept up in the crush are injured or killed. Weakly gregarious breeds include the Suffolk, Hampshire, Corriedale, Cheviot, Leicester, and Dorset. Because strongly gregarious breeds tend to move as a group instead of scattering, herding them is easier than mustering breeds that are not gregarious, especially when using a herding dog.

Biological Traits

Temperature: 102.5 degrees Fahrenheit

Pulse: 75 beats per minute

Respiration: 16 breaths per minute

Chromosome count: 54

Adult body weight: 65-475 pounds

Natural life span: 6-14 years (well-kept sheep have lived 20 years or more)

Sight: Although they have poor depth perception, sheep see in modified color and, unless wool on their faces obstructs their vision, they have a 270- to 320-degree visual field.

Smell: Sheep have a keen sense of smell. Rams determine which ewes are in standing heat by sniffing them; ewes identify new lambs by scent.

Hearing: Sheep are more sensitive to high-frequency sounds than we are; clanging gates and shrill whistles annoy and frighten them.

Teeth: Mature sheep have 32 teeth, including 4 pairs of lower incisors, but none in their upper front jaws; a hard dental pad replaces the absent upper incisors.

Advice from the Farm

RAMS AT A GLANCE

Our experts offer a few snapshots of rams, their behavior, and techniques for dealing with them.

The Crack of Heads for Dominance

"Rams start cracking heads as breeding season arrives—it's their way of showing their dominance.

"This summer, one of our young rams took aim on me a couple different times in a week's span. Each time, I responded with a swift kick. One time this fall, he was standing there looking at me from about ten feet away with a funny look in his eye. I kept my eye on him, and he walked up to me, placed his head on my thigh and pushed—he knew not to come at me from a distance and get up a head of steam.

"Rams tend to view us as part of the herd, too, and we have to keep our dominance in the flock to remain top dog. Otherwise, we are in trouble."

—Lyn Brown

The Dangerous Turn on a Dime

"Never trust them. Your sweet ram that has never hurt you once can turn on a dime and run you down. Be smart; keep your eyes on him."

—Laurie Andreacci

The Blaster Solution

"I have a bottle-raised ram that got a little aggressive. What we did was, we bought a set of water blasters. Every time he even suggested he was coming at us, he would get a blast in the face. It didn't take him long to figure out he'd rather stay away. After a while, we didn't need to carry the water blasters any more. About once a year, he starts acting rammy, and we have to have a refresher course."

—Lyn Brown

The Husband Toss

"Some people have a tendency to think that because rams are relatively small (compared to stallions and bulls), they are just sweet little things, as lambs are portrayed in so many ways. But these are animals with animal instincts, and they will try to dominate you—and often do.

"I had a large Suffolk ram keep me in a hay feeder we put out in the middle of the barnyard instead of close to 'people getaway places.' I was up there for about a half hour before he lost interest.

"We had an older North Country Cheviot ram who was the tamest, sweetest ram you could ever imagine. One day, after having not been in that particular pasture for some time, we herded them into the barn, and he rammed me against the fence. Before I could get up and untangled, he got me again. Luckily he couldn't get a good run at me, or he would have broken my hip.

"I've noticed that sometimes a good spray of water will stop them long enough to get away. I've even grabbed my husband (now ex) and put him in-between us to get away. It worked, too!"

—Connie Wheeler

The sheep leading the way along this road isn't necessarily the queen bee of the flock. She's just the first one to start walking.

While there usually is no flock leader, every flock of sheep does incorporate a pecking order, or social hierarchy. This is especially evident at feeding time, when high-ranking members eat first and those near the bottom eat last (if at all). In large flocks, several hierarchies may exist simultaneously. The ewes maintain a pecking order among themselves; rams maintain another; a third power play exists between the sexes. Jostling for position translates into head butting, shoving, and body slamming. Ewes are fairly subtle about it; rams indulge in all-out warfare.

Sheep want companionship, preferably that of other sheep. They'll make do with a goat, a pony, or a donkey companion, but they won't be as happy and might not thrive. They also prefer other sheep of the same breed. In mixed flocks, breed-specific subflocks are the rule.

When they can, sheep form family groups within their flock or subflock. A family might include an old ewe and her daughters and granddaughters, along with all three generations' suckling lambs.

Domestic versus Wild Sheep

- Most domestic sheep produce wool; wild sheep grow hair.
- Domestic sheep are generally shades of white or black; wild (and primitive domestic) sheep lean toward browns.
- Most domestic sheep have lop ears; wild sheep's ears stand erect.
- Some domestic breeds are naturally hornless; wild sheep, even ewes, are horned.
- Many domestic sheep breeds have more tail vertebrae than do wild sheep.
- The domestic sheep's brain is smaller than that of his ancient ancestors as well as that of today's wild sheep.

A flock groups together to face a large, aggressive dog. A few steps closer, and the dog will trigger the sheep's instinctive flight response.

STRESS AND FLIGHT ZONE

Because sheep are gregarious, separating them from other sheep stresses them. Harsh handling, loud noises, and the presence of predators (including dogs) also stresses them, as does lengthy confinement. Prolonged stress triggers cortisol production, and elevated cortisol levels significantly reduce immunity. Stressed sheep tend not to thrive. Symptoms include panting, restlessness, teeth grinding, and skittishness. Closely confined sheep, even those penned with their peers, gnaw wood and nosh on wool—their own or that of underling companions.

Sheep maintain personal security zones, or "space." Anything scary invading an individual's personal space generates flight. A sheep's flight zone might be fifty yards or nothing at all; breed, gender, tameness, training, and the degree of threat enter each equation.

If something arouses a sheep's suspicion, she will watch it closely. Her vigilance will alert others within the flock. If danger stays beyond the most timid observer's flight zone, the flock stays put. When her security zone is breached, she will turn and flee, and her flockmates will run then, too. Alarmed sheep stampede first and ask questions later. Instinct assures them that she who tarries is the sheep most likely to be eaten.

BREEDING TRAITS

Rams become sexually active between five and eight months of age. One mature ram can service thirty to thirty-five ewes in a typical sixty-day breeding season. Ram semen can be collected and successfully frozen. Ewes reach puberty between six and eight months of age, but they generally shouldn't be bred (depending on breed and maturity) until they're eight to twelve months old. Most ewes come into heat seasonally from early fall through early winter. Some breeds, such as Dorsets, Dorpers, Katahdins, and St. Croix, come into heat year-round.

This newborn sticks close by mama's side. Take time to observe the temperament of the parent sheep before purchasing a lamb. Jumpy moms produce spooked little ones.

A ewe's heat lasts twenty-four to thirty-six hours. Unless she's impregnated, she'll cycle every fourteen to nineteen days throughout the breeding season. Ovulation occurs twenty-four to thirty hours after the cycle begins. A ewe may be bred by natural cover (that is by a ram) or, more rarely, by artificial insemination, using fresh or frozen semen.

Most ewes lamb 145 to 155 days after conception, producing from one to as many as seven lambs, depending on age and breed. Twins and triplets are common. Ewes have only two teats, but so-called good milkers successfully rear triplets without help. Lambs are precocial, meaning they stand, nurse, and frisk about soon after birth. Commercial ewes are often culled when they reach five or six years of age, but depending on her breed and how she's been cared for, a ewe can often lamb until the age of twelve.

SHEEP CLASSIFICATIONS

Sheep can be classified by the types of wool or hair they grow, their shapes and sizes, and their regions of origin. Farmers also look at sheep in terms of dairy and meat production capabilities. Organizations such the American Livestock Breeds Conservancy think in terms of scarce and endangered breeds.

WOOL

You can judge sheep by their covers without a blush. Categories include fine and carpet wool, medium and long wool, and hair (instead of wool). British breeds, which have been exported to North America since colonial times, are categorized as either long-wool and luster or short-wool and down sheep. All are represented in North America today. Nearly every American breed incorporates British breed genetics.

A sheepish face looks out of a sea of wool. Length, texture, and luster of wool distinguish one breed of sheep from another.

Fine-wool sheep, such as Merino and Rambouillet (ram-boo-LAY), produce the soft, white wool used to make cushy, comfortable-next-to-the-skin garments. At the other end of the spectrum are the coarse-wool breeds, such as Scottish Blackface and Karakul, that produce the fleece used to stuff mattresses and to craft sturdy Scottish and Irish tweeds and rugs. In between are the medium- and long-wool sheep, whose fleeces are used mainly to make cozy outer garments and quality wool blankets.

Cheviots, Clun Forests, Dorsets, Finnsheep, Hampshires, Montadales, Oxfords, Polypays, Shropshires, Southdowns, Suffolks, Texels, and Tunises are medium-wool sheep. Long-wool and luster breeds, which produce long, lustrous, wavy, or ringleted fleeces of finer quality than those of their hill-bred brethren, include the Leicesters (LIS-ters) (Blueface Leicesters, Border Leicesters, and Leicester Longwools, Lincolns, Romneys, and Wensleydales). Other long-wool sheep are Coopworths, Cotswolds, and Perendales.

In Pullman, Washington, blackface ewes enjoy a crisp, autumn morning. The valley provides the perfect environmental conditions for the flock.

America's favorites, Suffolks and Hampshires, belong to the short-wool and down group. Less common (and in some cases endangered) short-wool and down breeds include the Clun Forest, Dorset, Jacob, Shetland, Shropshire, and Southdown. Some primitive sheep breeds grow hair instead of wool; others grow short, self-shedding fleeces. Because shearing can cost more than commercial wool is worth, and because hair-sheep lamb carcasses are bestsellers at ethnic meat markets, hair sheep, such as the Barbados Blackbelly, Dorper, Katahdin, St. Croix, and Wiltshire Horn (an ancient British breed), are a wave of the future.

SHAPE, SIZE, AND REGION

Other classifications of sheep deal with shape, size, and region. For instance, sheep tails tell their own tales. Fat-tailed sheep, native to the arid regions of Africa, the Middle East, and Asia, store body fat in their tails and rumps. This group includes Damara, Karakul, and Tunis sheep. Short-tailed breeds boast ancient Scandinavian roots, some via the sheep taken to Britain's northern isles by intrepid Viking settlers. Most are small, medium-wool breeds with soft, colorful fleeces and naturally short tails. They include Finnsheep, Icelandics, and Shetlands.

Classified by their small stature are miniature sheep breeds such as

My husband, John, carries Baabara, one of our Keyrrey-Shee miniature sheep. Minature sheep make the best of pets; a pair will be happy in a large backyard—and they'll keep it mowed and gently fertilized, too.

Babydoll Southdown and Keyrrey-Shee, which measure 24 inches or smaller at their newly shorn shoulders. American Brecknock Hill Cheviots measure 23 inches or less. Two other wee breeds are the Shetland and Black Welsh Mountain. Although miniature sheep are usually marketed as pets, they also sport fleeces that are a handspinner's delight.

Some breeds known by their regions are the mountain and hill sheep of Great Britain. These hardy, thrifty animals, generally medium size or smaller, adapted to life in Britain's rugged hills and highlands. Mountain and hill breeds found in North America include Cheviots (American Brecknock Hill Cheviot, Border Cheviot, Keyrrey-Shee, North Country Cheviot), Black Welsh Mountain, and Scottish Blackface.

DAIRY AND MEAT

Farmers raise sheep not only for wool but also for dairy products and meat. The American dairy sheep of choice is the high-producing East Friesian Milk Sheep. An increasing crowd of American sheep owners are clambering aboard the sheep-dairying bandwagon. Rich, mild-tasting sheep's milk is higher in protein, calcium, and fat—as well as vitamins A, D, and E—than is cow or goat milk, making it an ideal cheese-making medium. Many of the

Our ewe Angel is truly a rare breed; not only because she is a Wiltshire Horn crossbred, but also because she has horns (an unusual trait for ewes).

world's great cheeses are made from ewe's milk. These include Greek Feta, Kasseri, and Manouri; Spanish Iberico, Manchego, and Roncal; Sicilian Pepato and Ricotta Salata; and that perpetual French favorite, genuine Roquefort cheese.

Medium- and long-wool breeds are sometimes called dual-purpose sheep because they are raised for meat as well as wool. During the twentieth century, breeders created additional dual-purpose varieties by crossing Merino or Rambouillet fine-wool sheep with medium- and long-wool sheep. These new breeds include the Corriedale, Cormo, Romeldale, and Targhee. The British long-wool and luster sheep are also bred for both meat and wool.

SCARCE AND ENDANGERED

Commercial breeders generally raise big, meaty breeds, such as the Suffolk and Hampshire; wool factories prefer the Merino and Rambouillet. Many producers crossbreed for added productivity. As a result, a great many of yesterday's sheep breeds are becoming scarce and endangered.

The American Livestock Breeder's Conservancy publishes a priority list of threatened breeds. Critically endangered sheep include the California Variegated Mutant/Romeldale, Gulf Coast Native, Hog Island, and Santa Cruz. Breeds considered rare are the Cotswold, (American) Jacob, (American) Karakul, Leicester Longwool, Navajo-Churro, St. Croix, Tunis, and Wiltshire Horn.

CHAPTER TWO

Buying the Right Sheep for You

So you've decided you want to keep sheep. Although shepherding is far from complicated, it makes sense to learn all you can about sheep before buying any. Purchasing the sheep is just the first step. You'll need to find the best way to transport your new animals and to determine where they should be kept.

WHAT TO BUY

What sheep are best for you and your farm? Is it hot where you live? Cold, wet, arid? Are your pastures lush or rocky? Have you time to spend with your sheep, or would self-reliant sheep work best? Don't bring home stock before you're certain; know what you need before sheep shopping.

Of course, there's nothing wrong with bringing home a few "easy sheep" (whatever you can find locally) to gain sheep-keeping experience, as long as they're healthy and you enjoy them. But to save yourself a heap of heartache, do all the homework before launching a business or investing in pricey registered sheep.

CHOOSING THE BREEDS

At least five dozen sheep breeds are reasonably and readily available in North America, each developed to meet specific needs. Make a list of the qualities you're looking for, then go shopping. Perhaps you would enjoy basking in the glow of helping preserve a critically endangered breed (see sidebar).

Purebred, Crossbred, or Registered: Purebred sheep are those whose parents, grandparents, and beyond were all the same breed. Registered purebreds have papers to prove it. Purebreds "breed true," meaning their offspring inherit their parents' looks as well as many other traits, including elements of their dispositions and personalities.

Heritage Sheep

Want to keep sheep and also make a difference? Raise one of the endangered or rare breeds listed on the American Livestock Breeds Conservancy's Priority List.

The American Livestock Breeds Conservancy (ALBC) was founded in 1977 to preserve rare farm animals and promote genetic diversity in livestock. It acts as a clearinghouse for information on these subjects and registers breeds that don't have American registries of their own. It also assists in the conservation of heritage breed cattle, asses (donkeys), goats, horses, pigs, sheep, and poultry. The ALBC Priority List for sheep includes the following breeds.

- **Critical breeds:** Florida Cracker, Gulf Coast or Gulf Coast Native, Hog Island, Leicester Longwool, Romeldale/CVM, Santa Cruz
- **Threatened breeds:** Black Welsh Mountain, Chin Forset, Cotswold, Dorset Horn, Jacob-American, Karakul-American, Navajo-Churro, St. Croix
- **Watch breeds:** Lincoln, Oxford, Shropshire, Tunis
- **Recovering breeds:** Barbados Blackbelly, Shetland, Southdown, Wiltshire Horn

Sister organizations include Rare Breeds Canada and the Rare Breeds Survival Trust (UK). Australia, New Zealand, and many other countries support preservation efforts of their own. With their help—and yours—we can bring these endangered breeds back from the brink of extinction.

Crossbred sheep have parents of two different breeds; sometimes the parents are themselves crossbred. Because the parents are genetically dissimilar, crossbred sheep benefit from heterosis, or hybrid vigor, which translates into hardier, faster-growing animals. A crossbred often proves more productive than either parent.

If you plan to market breeding stock, to show at breed shows, or to participate in heritage or rare-breed preservation efforts, you'll need registered sheep. These sheep cost more—often much more—than do crossbreds or unregistered purebred sheep. In addition, although their breeding-quality offspring fetch higher prices than those of crossbred lambs, the market (meat) lambs and commercial wool of registered stock do not. Registered sheep take some of the guesswork out of breeding, especially for shepherds who know their breed's bloodlines and how best to use them. Registered flocks are generally more uniform looking than crossbred or grade (unregistered) herds. If this is important to you, buy registered sheep.

If you just want productive sheep, think crossbreds. Good ones combine the best of their parents' breeds, plus the punch of hybrid vigor, in a single, less costly package. Often they're also more readily available than purebred or registered sheep.

Availability: When choosing a breed, consider availability. You could opt for a

This handsome ewe's sire was a Scottish Blackface ram, her dam a Black Welsh Mountain ewe. Crossbreds often inherit the best qualities of both parents.

breed common to your area (in which case you wouldn't have to travel far to purchase seed stock or replacements) or something unique (thereby making yourself the only regional source of stock for others who might like to own this breed).

If you decide on something unusual, you'll have to buy your foundation animals from afar. Although transportation may be an issue, keep in mind that horse and livestock transporters will haul sheep for a fee and that lambs of miniature sheep, such as Olde English Babydoll Southdowns, Keyrrey-Shee, American Brecknock Hill Cheviots, and even some Shetlands can be shipped between major airports in maxi-size dog crates. However, a single sheep is a stressed sheep; when shipping sheep, always buy multiples if you can.

For locally common breeds and crossbreds, a number of sources exist. Watch newspaper classifieds and ask at feed stores, veterinary practices, and county extension services throughout your search area. In addition, you can peruse registry directories and breed journal ads.

Don't buy your first sheep at auction! Livestock sales are the farmer's dumping

Did Ewe Know?

Many of today's sheep breeds have long, illustrious histories. These include the Border Cheviot (first recognized in 1372), Shetland (taken to the Shetland Islands by Viking settlers), Tunis (pre-Christian era), Karakul (pictured in Babylonian temple carvings), Navajo-Churro (brought to the New World from Spain circa 1493), Welsh Black Mountain (pre-Middle Ages), Cotswold (pre-Roman conquest), Icelandic (carried to Iceland by Viking settlers in the ninth and tenth centuries AD), and Jacob (possibly mentioned in the Bible; raised in Britain for more than 350 years).

ground. Except for slaughter lambs, most sheep sold at auction are culls: aged ewes with broken teeth, ornery rams, and sheep afflicted with hoof rot (or worse). Even healthy sheep passing through sale barns are exposed to sheep that are not. Experienced shepherds occasionally bag a jewel in the rough, but until you can recognize the many signs of potential pitfalls, you're better off purchasing sheep at private treaty from a breeder.

When you do buy, buy two. Because of its gregarious nature, a solitary sheep will be frightened and lonely. One possible exception is a bottle lamb (or adult former bottle lamb) that accepts humans as its flock.

Talking to Breeders: When you've narrowed your choices to a few favorite breeds or crossbred types, contact registries or peruse online directories to find breeders in your area. Visit as many as you can. Shepherds love to talk sheep, and you'll learn worlds by picking their brains.

Producers' Web sites are another great source of valuable information. Find them via registry Web sites, search engines, and online breeders' directories. In addition, breed-specific books and journals are a godsend for newbie shepherds, as well as a means of uncovering bloodline data and health-issue information you might not find elsewhere (see Resources).

The best way to obtain information—either general or specific—about sheep is to work with a mentor. If a local shepherd befriends you, consider your-

Though she is in the background, this brockle-faced ewe stands out from the group. Uniqueness may be a selling point when shopping for sheep, but carefully consider special requirements before making a decision.

self blessed. Barring that, you can meet loads of helpful, experienced shepherds via the Internet. Don't be shy! Ask one or more friendly souls to coach you; most will be happy to oblige.

SELECTING THE SHEEP

You've selected your breed and contacted a breeder. Now you're off to see some sheep. When assessing the sheep, you must consider several factors, including conformation, health, and teeth, as well as sex-specific qualities, such as udder condition.

Conformation: When evaluating purebred sheep, assess their type. Type is what makes sheep of a given breed resemble one another. It's outlined in each breed's standard (a registry-generated guide to that breed's ideal). If you understand your breed's standard before you go shopping, you'll select better sheep.

Quality sheep of all breeds have traits in common. You'll want straight, sturdy legs, set one in each corner; such sheep have wide, strong hindquarters and broad chests. They have deep bodies, well-sprung ribs, level backs, and fairly straight underlines. They're wide and firm at the haunches, and their backs are nicely fleshed. If fleece conceals their basic structure, explore prospective purchases with your hands—that's the only way to know for sure what's under there.

Health: Any sheep you buy should be healthy. Healthy sheep are alert and interested in their surroundings. Unhandled sheep will be on guard (but not frantic), and tame ones may beg for attention. Avoid droopy, disinterested sheep that cough, wheeze, or just plain strike you as sickly. Healthy sheep have reasonably tidy backsides, non-drippy eyes and noses, and bright pink mucous membranes. Raggedy, moth-eaten

A Scottish Blackface sheep grazes happily in the midday sun. To ensure health as well as happiness, apply a hands-on approach to evaluating prospective additions to your flock.

fleeces bespeak sheep riddled with lice, mites, or wingless biting flies called keds. Limping sheep may be suffering from a dreaded, contagious disease called foot rot. Unless you know for certain they're not afflicted with such a malady, don't buy them!

Teeth: A sheep has a pad of hard tissue, but no front teeth, in its upper front palate; eight incisors grace the front lower jaw. Lambs grow eight baby incisors, which are replaced by two larger permanent teeth each year, beginning at the age of one. They start at the center and continue outward, so you can easily tell a sheep's age through year four. Once they've emerged, a sheep's teeth continue to wear and spread farther apart; by the time they're eight or nine, most sheep have lost or broken some of their incisors. Because they can no longer efficiently graze, older animals require specialized feeding; don't pay top dollar for "broken mouthed" or "gummer" (toothless) sheep.

By the same token, unless a sheep's lower teeth align properly with its hard dental palate, she can't efficiently grasp and rip off grass. If her upper palate juts out farther than her teeth, she is "parrot mouthed"; if her incisors protrude beyond her dental pad, she's "monkey mouthed." Either way, she's a problem feeder and a sheep you should probably avoid.

Sex-Specific Factors: When buying a ewe, inspect her udder. You can reach under her to do this or ask her owner to set her up on her hindquarters, which is better still. She should have two symmetrical, widely spaced teats set on a warm, soft bag. Sheep shearers in a rush sometimes zip off teats. Lumpy udders are a mark of mastitis. You might have to bottle feed a problem ewe's lambs; if that's not an option, refuse her.

If you aren't lambing savvy or have chosen a breed known for lambing problems, pick experienced ewes. Ewes reach peak productivity between four

Our ram Abram pants at the sight of the treat bucket. His permanent teeth recently replaced his baby incisors; as he ages, his teeth will widen and then wear down. As a toothless senior, he'll need a special diet.

and six years of age. However, large-scale breeders often cull six- to eight-year-old ewes that can work perfectly in a hobby farm setting. They'll teach you loads in exchange for a smidgen of extra care.

When buying pets, forgo the rams. Rams are unpredictable; a surly one is trouble on the hoof. If you can't resist that winsome ram lamb, think castration. Altered males (called wethers) make wonderful pets. However, if you need a breeding ram, buy the best one you can afford. He'll influence your entire lamb crop, so he should exem-plify the best of his breed. He must have two large, smooth, roundish testicles suspended in a free-hanging scrotum; never buy a ram with a single testicle. In sheep, size counts. Measure a ram's scrotum at its widest point. Miniature and small-breed rams should tape close to twelve inches, and large ones, four-teen to fifteen inches or more.

If he's horned, make certain you can place two fingers horizontally between his face and horns. Sheep chew with a side-to-side motion; if his horns crowd his jaw, he could starve.

Advice from the Farm

CHOOSING AND BUYING SHEEP

When choosing a breed or buying sheep, keep in mind these breed selection tips provided by our panel of sheep experts.

Read up on it

"(When I selected my breeds), Ipurchased a book off Amazon.com titled In Sheep's Clothing. This book listed every breed available, along with their temperament, environment they do well in—damp versus dry, green feed versus dry land. It also listed their expected wool clip per year, mature size for each sex, micron count for the wool, and lambing percentages. Also, some of the best information out there is available from your state extension service."

—Lynn Wilkins

Talk to Old Timers

"Talk to old-time sheep buyers if you can. They have worked all classes of sheep for years and know just about all there is to know about these animals. They can tell you which sheep are most likely to be a profit for you, and this is important."

—Connie Wheeler

Do Your Homework

"Do your homework. Locate a reputable sheep breeder, then tell the breeder what you are looking for. Let them select the animals they feel will work for you. I was raised on a ranch and have been around animals all my life, but I bought my present sheep sight unseen from breeders who I researched via Web sites and chat lists. I knew they were top-level breeders and were producing the quality of sheep I wanted. They were also the breeders who welcomed my questions via e-mail and took the time to respond. And I joined all the Yahoo groups that discussed the breeds I was considering."

—Lynn Wilkins

Get What You Want

"If you really like a breed, get it. I think if you care for your animals, as in feeding and health care, then they'll do well no matter where you are located. It might cost you a bit more to have them shipped to you, but you'll have gotten the breed you really wanted, and in the end, you'll take care of them better for it."

—Laurie Andreacci

Help Each Other

"Never be afraid to ask questions. And remember, we're all in this together, so help out your fellow sheep breeders."

—Laurie Andreacci

This Scottish Blackface ram's face is swollen between his eyes and nose, a normal occurrence when breeding season approaches.

The Sale: When buying expensive, registered stock, request references (and check them out). In every case, ask for detailed health and production records. If you'll be crossing state lines with your purchases, arrange for health papers (although you'll probably have to foot this bill). Read those registration papers carefully. Are they up-to-date and transferred to the present seller's name? Do identification numbers match those tattooed on sheep or printed on ear tags? Make sure you get the right sheep!

Beginning soon, all sheep breeders will be required to participate in one of two federal scrapie-eradication programs. All sexually intact sheep and all wethers more than eighteen months of age will be permanently ear tagged, tattooed, or "microchipped" with their breeder's premise number before leaving his or her premises. These are government-mandated programs, so don't buy sheep from a noncompliant herd!

The sheep's ear tag indicates the breeder's premise number. Tagging is one method of compliance with government-mandated livestock identification in an effort to control spread of disease.

EWE HAUL . . . AND THEN?

Hauling your sheep can be as simple as boosting them into the topper-sheathed back of your pickup or as complex as hiring an animal transporter to truck them across the country to your farm. To make the move as smooth as possible, do everything you can to make your sheep comfortable. Bed their conveyance so they can lie down en route. Don't use sawdust, wood chips, or finely chopped straw; all work their way deep into fleeces, reducing their value. Long-stem straw or grass hay works well; it's soft and cushiony, and sheep can nibble it if they want. On long trips, stop and provide drinking water every few hours. Ensure that the conveyance is a safe one. Sheep can jump higher and push harder than many people imagine.

Sheep don't tolerate stress. Keep things low key, and—unless it's absolutely necessary—don't haul a sheep by itself. Conversely, don't crowd sheep into a space that's too small. Sheep need room to move around. Ensure adequate ventilation, without drafts.

Holistic shepherds sometimes dose each sheep with probiotic paste or gel before and after hauling. This boosts immunity and reduces tummy upsets. In addition, spritzing their faces with Bach Rescue Remedy will help steady your sheep's nerves.

Have your facilities ready to receive the sheep on arrival. This includes laying in a supply of the same sort of feed their former owner fed them. Switch feeds gradually to prevent nasty digestive upsets.

New sheep should be quarantined for at least three weeks. They should be dewormed—always. If their vaccination history is uncertain, revaccinate. When quarantine time is up, introduce new ewes gradually, perhaps by penning them in an enclosure adjoining the existing flock's pasture. Rams are another story.

A simple enclosure such as this will work well for quarantining new sheep from the rest of your flock. It's also useful for separating ewes from lambs at weaning time.

CHAPTER THREE

Housing, Feeding, and Guarding Your New Flock

Housing for hobby farm sheep is the essence of simplicity. Sheep are happier and healthier when kept outdoors; you need to provide your flock with well-maintained pastureland. Yet they will also need fencing to keep them safe from predators, simple shelters to protect them from the elements, and pens and stalls for lambing and times of quarantine. Wherever you keep your sheep, you'll need to make sure they have all the amenities.

Fencing and shelters alone often are not enough to adequately protect your flock. For additional protection, you should consider engaging the help of a guardian animal, whether that be a specially trained dog, a llama, or even a donkey.

LIE DOWN IN GREEN PASTURES

America is a huge country; Minnesota pasture species don't thrive in the Ozarks, nor do western forage grasses in the Deep South. Again, touch base with your friendly county agricultural extension agent. He can help you choose what's right for your locale and teach you how to plant and maintain it.

Provide plenty of water in your pasture, and keep it sparkling clean. Sheep won't drink enough to satisfy their needs if their water supply is alive with green algae or is a scummy mess the flock has pooped in. Don't just dump and refill the trough; scrub it out. A weak chlorine bleach solution helps chase away stubborn algae deposits. Feed stores sell low, sheep-appropriate fiberglass and metal water troughs, but we prefer injection-molded plastic horse stall cleaning baskets. Heavy-duty plastic cattle and horse mineral lick tubs and pans make dandy, easily cleaned sheep waterers, too. Pastured sheep need shelter from sun, hail, and possibly snow and freezing

Lush, green pasture provides essential nutrition for a grazing flock. Given their choice of victuals, sheep choose 40 percent grass and 60 percent browse, meaning they'll cheerfully rid your meadows and open woodlots of tender tree sprouts, brambles, and briar roses.

rain. Trees work fine as sunshades, but your sheep need manmade shelter to withstand harsh storms. If you don't take your sheep inside at night, make a point of checking them at least twice a day. Count noses. Walk around each sheep, checking for illness or injuries. Patrol all fences and associated structures on a weekly basis, scouting for downed wire, exposed nails, loose tin, wasp nests, and any other accidents waiting to happen.

Did Ewe Know?

In days of yore, shepherds in Europe were often buried with a tuft of wool in their hands. This practice sometimes signified their devotion to their charges. At other times, it simply showed that they were shepherds—thus excusing occasional lapses in church attendance, because they couldn't leave their flocks during lambing.

Provide pastured animals with a high-quality, free-choice, granulated sheep mineral mix. Place it inside their shelter, or buy or make a covered container to install outside. Feed a little "sheep candy" every day. A scant handful of corn, pellets, or sweet feed per head keeps them happy to see you, and happy sheep are a whole lot easier to handle.

If your sheep don't keep weeds and saplings in check, occasionally mow the pasture—or buy a few goats to add to the flock. Constantly grub poisonous plants out of your pastures. Many are bitter, and well-fed sheep won't eat them—although there are notable exceptions. It takes only a nibble of certain species to dispatch a lamb. It's best to err on the side of caution.

Our lamb Ewephemia calls to her mother from inside a sturdy, well-ventilated enclosure. The fence is key to the weaning process. Lured in with grain, the flock is released from the enclosure one by one until only the newly weaned lambs remain.

(Do) Fence Me In

In most cases, a sheep fence is more important for keeping things out than for keeping them in. Sheep are incredibly easy to fence; one nose-first encounter with electric wire convinces them they want to stay put. Keeping Arkansas coyotes and other predators at bay is another story.

One solution is to coyote-proof nighttime quarters for your sheep. Come sundown, they'll be in a nighttime fold, with tightly stretched, all-the-way-to-the-ground, woven wire fence surrounding the area. You can reinforce it with strands of electric fence along the top and outside bottom. Should you leave the farm during the day, the sheep can go in then, too. Dogs can be far more lethal than coyotes, and dogs tend to strike during daylight hours. You can purchase a goat companion or two wearing bells to warn of approaching danger.

Some variables to consider when selecting fencing are soil base (for example, rock, sand, or clay), lay of the land, proximity to roadways or neighbors, breed(s) of sheep involved and associated predators from which you want to protect them, available fence-erecting labor, and fencing costs. Fat books have been written about fencing; it's beyond the scope of this book to discuss the hundreds of types of fencing available in the United States. Before deciding on one, talk to your county agricultural agent and peruse the fine agricultural bulletins available online from the university

Advice from the Farm

HOUSING AND FENCES
Our expert sheep advisers talk about housing and fences for sheep.

Housing Hair Sheep
"Although we do have barns and other sheltered areas, most of our sheep don't use them except on extremely rainy, windy days. They'll stay out in the rain most of the time. Hair sheep have a thicker skin to compensate for their lack of wool."

—Connie Wheeler

A Good Pasture
"It's good to have a hilly pasture where your sheep get lots of exercise going up and down hills to feed and water."

—Connie Wheeler

Close the Gate!
"My mantra is, 'Always close all of the gates, all they way, all the time.' Even if you're just popping into the pasture to grab a bucket or something. It's tempting to not always bother. Sometimes you get away with it, but sometimes not."

—Lisa MacIver

The Ideal Fence
"The ideal fence would be made of the kind of non-climb fencing that equestrians use, although this is quite expensive compared to regular 'field fence.'

"An electric wire placed away from the fence about twelve inches and off the ground about twelve inches will keep sheep completely away from the fence. It's also a good idea to spray vegetation killer under and on both sides."

—Connie Wheeler

Designed to house miniature horses, this inexpensive, three-sided plywood structure with attached pipe gate enclosure provides adequate shelter for sheep.

resources listed in the Appendix of this book. Send for the Premier fencing catalog (contact information also provided in Appendix). If it's about fencing, it's in there: which types work where, what each costs, and exactly how to erect it. The energizer (electric fencer) section is an education in itself. If you aren't already fencing savvy—or even if you are—get the catalog!

SHELTERS

All a small flock (twenty-five or fewer sheep) really needs is access to simple shelters such as the one pictured here, erected with their open sides facing away from prevailing winter winds. They're inexpensive and portable, useful for cleaning purposes, and they can be simply constructed by any reasonably adept home carpenter in a day. Be sure to allot enough space in your shel-ters for the number and type of sheep you own. Eight square feet is right for most breeds; twelve square feet is adequate for ewes with lambs. We call these structures mini-folds in honor of our Cheviots' Scottish roots.

Whenever possible, place shelters on a southern slope with well-drained soil. Keep them close to your house to discourage predators and to make midnight-lambing time checks a little bit easier on a sleepy shepherd. Available space in an existing barn or shed can be used as well; just remember that sheep need to get outside at least part of the day. Indoor accommodations must be well ventilated but draft free.

PENS AND STALLS

You'll need an indoor spot to erect your lambing jugs (small pens where ewes are confined with their newborn lambs for a few days postpartum; we'll talk more about jug requirements in Chapter 6) and a quarantine, or "sick sheep," stall. The areas needn't be elaborate, just big enough for a single sheep or a ewe and her lambs to be comfy, dry, and out of drafts.

When choosing bedding, seek absorbency in a form unlikely to work its way deep into fleece and thereby spoil it. Don't bed shelters, jugs, or quarantine stalls with wood shavings or saw dust. Shavings quickly imbed themselves in fleeces and ruin them, and the fine dust particles in sawdust are hard on sheep's respiratory systems. Dust-free long-stem wheat straw is the bedding of choice, but other types of long-stem straws are all right, too. Avoid finely chopped straws of any kind. Other workable options include peanut hulls, ground corncobs, and the hay your sheep inevitably pull down and waste.

Can Ewe Safely Eat It?

There are few things more heartbreaking than losing animals because they ate poisonous plants that you should have grubbed out of their pasture. As shepherds, it's our obligation to make pastures as safe as possible for our woolly friends. Unfortunately, it can be hard to know which plants and brush to remove. What's perfectly safe for one livestock species to browse can dispatch another in record time, and flora varies widely from place to place, so contact your county agricultural agent to identify mystery plants your sheep are munching. Don't assume they're all safe to eat.

SHEEP FEEDING PRINCIPLES

Only a few classes of sheep routinely require concentrates: ewes during flushing, late-gestation and lactating ewes, hard-working rams, geriatric or underweight sheep, and an occasional lamb. Concentrates must be sheep specific: whole or cracked grains, or commercial products labeled for sheep. Although the subject is complicated, suffice it to say sheep are easily poisoned by too much copper in their diets. Swine and poultry feeds are high in copper, as are some beef and dairy cattle rations. Even some goat mixes contain more copper than some breeds can handle. It pays to look for the words "sheep feed" on the label. For the same reason, you shouldn't expose sheep to mineral licks or loose minerals designed for other species. Sheep require minerals, but in sheep-specific form.

No matter what some folks insist, sheep cannot safely digest spoiled or moldy feed, including musty hay. Consider mycotoxins—poisonous compounds produced by molds. They could be in that moldy feed. Moldy feed can trigger bloat, too. Measure feed, especially concentrates. Don't feed a lot today and half that amount tomorrow. When group feeding concentrates, don't let bossy gluttons hog it all. They can bloat or get acidosis and founder, while meeker flockmates literally starve. Sheep are creatures of habit. Try to feed them about the same time every day. They prefer to eat during daylight hours and in the company of other sheep. If you must switch feeds, do so gradually, over a ten- to fourteen-day period to allow sheep's rumens to adjust.

HAY THERE!

Forage is the basis of most sheep diets. Consider hay, for example. Most hobby farmers have to buy it; you're exceedingly lucky if you don't. In some parts of the country, acquiring decent hay is an expensive proposition. It's either not locally grown or not fit to feed. If you live in a place where quality hay is grown, you are at an advantage.

Small square bales usually work best for sheep; they're easier to handle and store in a hobby-farm situation. Because you can dole out exactly what they need, the sheep waste less when fed from smaller bales. Large round bales are acceptable if they've been stored under cover and are free of mold. However, your sheep will pull hay down, then defecate and lie in it. You also can expect a lot of debris-contaminated fleece at shearing time. Nutrient-rich hay is green. Quality alfalfa is dark green; if it's a strange lime green, it has been treated with propionic acid preservative. Authorities claim it's safe, but if you're not convinced, avoid it. Well-stored grass hay is light to medium green. The outsides of both legume and grass hays become bleached if exposed to light in storage, but inside the bales the hay should still be green.

If you see white dust, blemishes, or black spots, or if there is a musty smell, the hay was cut and stored before being fully dry. Even if you don't spot actual

Hay bales of this size are a good choice for a hobby farm. Depending on the number of sheep you keep, larger sized bales sit longer in storage, gradually losing certain nutrients.

Golden yellow hay such as this offers up a nice snapshot, but it is likely to make poor feed for sheep. When purchasing bales, examine them closely for color, smell, texture, and moisture.

mold, this hay shouldn't be fed. Hay that's pale to golden yellow inside and out wasn't put up promptly; it got sun bleached in the field. It's nutrition poor and apt to be dusty. Furthermore, it's definitely not a good buy.

Quality hay is made from leafy plants mown just before blooming. It has thin, flexible stems that easily bend without breaking. Coarse, stemmy hay is low in nutrients and unpalatable, too. Leaves contain most of the plant's protein; stems are mainly hard-to-digest, low-energy cellulose. As plants mature, they grow more stem than leaf. Stemmy hay made from older plants is no bargain, and hay containing mature blossoms and seed heads was mown well past its prime.

Refuse hay containing sticks, rocks, dried leaves, weeds, and insect or animal parts: your sheep can't eat the sticks and rocks; the weeds may be toxic, and

their seeds will infest your property; and baled insects and desiccated animals can cause serious disease. Don't feed your animals uncured new crop hay, either. Age new hay fresh from the field for five to six weeks before feeding it. If you haven't enough hay to last, buy enough dry hay to tide you over. Feeding uncured hay can easily trigger bloat. Old hay isn't necessarily bad hay. Nutritious hay put up dry and carefully stored holds its nutrient content for several years. When its still green inside, old hay can be a bargain.

Whenever possible, buy tested hay from a reputable hay dealer. It takes a lot of guesswork out of the equation. Be extra cautious if you buy hay at auction. Some sellers place their best bales on the outside, where buyers can see them, and tuck junky hay deep inside the stack. Inspect the hay you plan to buy. Agree to pay for a couple of bales you can open. See what they look like inside. If the hay doesn't pass muster, you aren't obligated to buy the rest.

If you can have your hay delivered and neatly stacked in the barn, do it. Unless you're young and energetic, and have a great back and knees, it's extra money well spent. Buy only the amount of hay you can store under cover. Stack it on pallets to get it off the floor; if you don't, the bottom layer will rot. Don't let barn cats defecate or have kittens on it; cats are primary carriers of toxoplasmosis. Finally, to minimize waste and prevent disease, use those feeders. Never feed hay off the ground.

ALL THE AMENITIES

Your sheep need something to eat out of. There are countless styles of hay and grain feeders to choose from. Large, sturdy-plastic grain feed pans are tough, portable, and easily cleaned. Wall-mounted steel rod racks work well for hay. A great portable hay feeder for just a few sheep is a horse-style nylon hay bag. You'll find durable multisheep feeders at farm and feed stores, or you can order them from sheep supply catalogs.

Another option is to build your own feeders. If you request item #900000 when placing an order from the Premier sheep supply catalog, the company will send free plans for its walk-through and drive-by flock feeders. The walk-through plan is especially nice if your flock includes a testy ram.

Sheep shouldn't eat out of standard big-bale feeders. When they bury their heads in the hay to choose the choice morsels, they imbed a lot of nasty debris in their fleeces. They may also rub off patches of their fleeces when they lean against pipe rails while eating.

- Don't feed hay off the ground. Once one sheep defecates or urinates on it (and one will) none of the others will eat it. Eating off the same stretch of worm-egg-infested earth encourages worm infestation, too.
- Mount feeders reasonably close to the ground so hay chaff doesn't drift into faces and fleeces while sheep are eating.
- Locate feeders where you can place hay or grain in them rather than flinging it. Sheep raise their heads when it's least expected: voilà, a face full of hay or grain.

Two of our girls, Ewephemia and Ewelanda, munch supper from a secondhand cattle mineral lick pan. Feed stores carry myriad styles of livestock feeding and watering troughs. Alternatively, you can be creative with homemade or used versions.

 # Advice from the Farm

SHEEP FEEDING TIPS AND TRICKS
Our panel of sheep experts discuss feeds and feeding.

Feeding Geriatric Sheep

"After breeding time, we take our older girls and pen them together in a separate area from our younger ewes. We call it our geriatric pen. These old ewes get about one-half pound each per day of Equine Senior. It has made an amazing difference in their weight, activity, and the babies they produce. For a treat, they love bread."

—Lyn Brown

Selecting Feed

"My four kids raised the three main species of farm animals for 4-H and Future Farmers of America (FFA) market shows, with numerous grand champions along the way. We always fed the feed that the local feed mill produced for that particular species and our specific location . It worked well for us. The feed contained the trace minerals our area needed and the feed composition percentages the species needed."

—Lynn Wilkins

GUARDING YOUR SHEEP

According to the National Agricultural Statistics Service, predators (coyotes, dogs, bears, cougars, bobcats, foxes, and eagles) killed nearly 5 percent of America's sheep population in 1987. This represented a loss of $83 million to the farmers and ranchers who kept them. By 2009, predation losses were down to $20.5 million. The reason? A growing legion of sheep producers have been turning to flock guardians—dogs, donkeys, and llamas—to protect their flocks.

Canine flock guardians are not herding dogs or pets; they're specialists bred to guard livestock—and they do it well. Originating in Europe and Asia, guardian breeds include the Great Pyrenees (from France), the Yugoslavian Sarplaninac, the Hungarian Komondor and Kuvasz, the Turkish Akbash and Anatolian Shepherd, and the Italian Maremma Sheepdog. These dogs are raised with and bond with sheep, so they simply remain with their "flockmates" at pasture. About three-fourths of these dogs become good guardians by actively repelling small predators and by alerting their owners when the flock is threatened. However, some simply don't like sheep, others wander, and still others want to stay by the house and be pets.

Guardian donkeys live longer than do guardian dogs, they're inexpensive to buy and maintain, and they require no special training. Like llamas, donkeys instinctively dislike the canine clan and protect their herds (be they sheep, goats, poultry, or other donkeys) from stray dogs, coyotes, and even wolves. A donkey in attack mode can gallop and stomp at the same time. Needless to say, a new guardian donkey must be carefully introduced to family dogs—who will quickly learn to give the newcomer a wide berth! Like dogs, some donkeys dislike sheep and will hang around the barn seeking human contact instead of minding the flock. However, people who like donkeys swear by these long-eared guardians.

Llamas may be the best bet for hobby flock guardians. Llamas bond with their woolly charges extremely well, eat the same feed, and require most of the same vaccinations as sheep. Llamas, especially females with cria (suckling offspring) and castrated males, instinctively guard their herdmates, even when their "herd" consists of sheep. Guard llamas shriek at, charge, and sometimes stomp marauders. Few small predators (such as dogs and coyotes) willingly tackle the same angry guard llama twice. (See the Appendix for more information.)

This handsome Great Pyrenees guards the farm of Dick Harward and Barb Zebbs. Great Pyrenees are the North American livestock guardian dogs of choice. Other favorites include the Akbash, Anatolian Shepherd, and the Tibetan Mastiff.

CHAPTER FOUR

Sheepish Behavior and Safe Handling

Most people think sheep are stupid, but they're wrong. Folks who keep them tell of clever ewes teaching lambs to split chestnut hulls with their hooves or a flock banding together to shake apple trees to harvest succulent fruit. Sheep are good at sheepy things but sometimes not as adept at performing tasks their caretakers prefer. For example, a sheep may not happily leave her peers and move quietly from place to place in an orderly fashion. The whole flock may not thrive indoors in close confinement. These things are foreign to the inner clock that makes sheep flocks tick.

Shepherds who have difficulties working with sheep have never learned to think like their charges. These animals operate by an ingrained set of rules that are rarely broken. When we learn to play by sheep's rules, sheep and humans exist in perfect harmony. People who don't learn the rules tend to call sheep stupid. Those who understand how sheep communicate and how they perceive and interact with their world find sheep intelligent, ingenious, and easy to handle. We know you want to join the latter camp.

Your first step is to learn what factors have shaped sheep instinct and how sheep typically behave as a result. Next, learn to read your sheep, who communicate through intricate verbal and body signals. This knowledge will help you achieve your goal of handling your sheep effectively.

WHY THEY DO THAT THING THEY DO

Sheep know they are vulnerable to predators. They are small and scrumptious creatures who don't bite or kick to defend themselves, and they rarely use their horns in self-defense. So they are supremely vigilant and suspicious. When star-

Following centuries of instinctive programming, a large flock keeps close together while grazing. Just as their ancestors were, the sheep are vulnerable to predators who also make their homes in these beautiful hills.

tled, they are wired to flock and run. Historically, Mother Nature culled bold sheep and any others that paused to ponder. Timid, wary sheep bred on.

Although wily coyotes and sheep-munching wolves aren't fixtures on today's hobby farms, sheep remain unconvinced of their relative safety. They know it's easier for a wolf to snag a solitary sheep than to pick one out of a tightly packed, fleeing flock, so they first seek safety in numbers. Though flocking behavior does vary by breed—from supremely gregarious white-faced sheep, such as Merinos and Rambouillets, to self-reliant Cheviots, Leicesters, and most hair-sheep breeds—all sheep flock to some degree.

FLOCKING

Flock dynamics apply to mobs of four or more sheep; fewer sheep may not respond as expected. Dr. Ron Kilgour, of New Zealand's Ruakura Animal Research Station, assessed flock dynamics by working sheep in a maze. Single sheep were considerably stressed but eventually mastered the maze quite well. Two sheep went in different directions, and three sheep scattered; both of these groups generally muddled things up. Four or more sheep moved as one and learned faster than single sheep. That's why three sheep are the norm at herding trials; they give herding dogs the acid test.

> ### Did Ewe Know?
>
> Sheep on the West Yorkshire moors have learned to roll on their sides, commando-style, across eight-foot cattle grids, guarding the town of Huddersfield. Of them, a National Sheep Association spokeswoman said, "Sheep are quite intelligent creatures and have more brainpower than people are willing to give them credit for."

Because of their natural flocking instinct, a number of sheep may be skillfully herded together in a single direction, like these. If they are frightened, however, their fleeing instinct will transform the herd into a mob.

Sheep follow a leader. The leader might be a shepherd with a bucket of grain, but it's more likely another sheep. Leaders aren't necessarily the brightest nor the highest-ranking sheep in the flock. They're simply the first to start off in a given direction.

FLEEING

Nervous sheep mob together, then usually flee. If terrified, they flee blindly. If sheep in the front of a flock ram into a barrier, there can be pileups as rear runners tailgate flockmates directly in front of them. Frightened sheep don't "do" sharp corners, including the space behind swinging gates. They pack into them and stay there.

Calm sheep move forward at a walk, and frightened ones stampede or back up. Calm sheep are thinking sheep; sheep in a panic just react. When her flight zone is penetrated by a perceived predator (including a dog or shepherd), a sheep moves away.

Sheep move forward out of darkness, toward light. Their faulty depth perception renders shadows and harsh light/dark situations terribly frightening. They move from confinement toward open spaces, into the wind rather than downwind, and more readily uphill than down. Sheep prefer not to cross water, they dislike passing through narrow openings, and they panic on slippery surfaces, be they natural or manmade (such as wet stall mats and slick concrete). Sheep have long memories. Most of them easily recall bad experiences for a year or more.

DO EWE READ ME?

Sheep's verbal and body signals vary somewhat depending on the number of sheep in the flock and its members' sexes, ages, and breeds. Skilled shepherds easily "read" sheep, and you can, too. The trick is to spend time among your sheep, simply hanging out and observing. Sheep will teach you a lot that way.

READER POSITIONS

Sheep are more at ease with humans who get down on their level. Consider squatting or upending a bucket and sitting among your sheep to observe them. Don't do this if your ram is aggressive, and don't stare directly into a dominant sheep's eyes. By the same token, if confronted by a testy sheep, you should try to look big. Stand tall and hold your arms up and out to the sides. If you spy anything to extend your outward reach—such as a stick, shepherd's crook, or hand tool—use it.

SHEEP SIGNS

Once in a safe position to read the sheep signs, look and listen for some of the following:

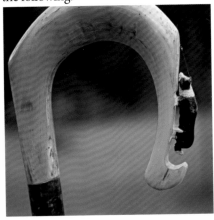

This traditional crook-style staff was carved from a Scottish Blackface horn. Dick Harward and Barb Zebbs of Wind Fall Farms craft these and other sheepy treasures in Willow Springs, Missouri.

Ear movement. Sheep depend greatly on their keen hearing. Breeds with upright ears are often more reactive than are floppy-eared sheep, simply because they hear scary things the others miss. A sheep with erect ears is more easily read than others. If her ears are pricked alertly forward, she's focused on whatever she's looking at, or she may be intently listening. Ears intently upright but swiveled toward the rear mean she's listening to something behind her; she'll probably turn and face that direction, the better to see and hear whatever's captured her attention. Pinned ears (ears laid back close to the head) generally signal annoyance or aggression but may indicate fear or excitement. A relaxed sheep's ears face to the sides; she's not focused on anything in particular.

Stamping, nodding, glaring. Annoyed sheep may stamp their hooves, raise or nod their heads, or glare. They usually back a few paces before charging, although ewes sometimes bash each other from a standstill. Rams or extremely pushy ewes may rub or bump humans with their foreheads; this is a sign of early aggression and should be strongly discouraged. When working among sheep, the wise shepherd avoids unnecessary handling of foreheads, foot stomping, or walking backward, which can all be misconstrued by sheep as inviting aggression.

Panting. A ram pursuing his heart's desire pants like a freight train. He grumbles at his intended in a low, seductive tone, sniffs her backside, and rolls his upper lip back in the flehmen response. This distinctive grimace blocks his normal breathing apparatus and draws air through a pouch-like opening in his upper palate called the Jacobson's organ. By analyzing her scent, he determines if the ewe he's courting is in heat. If she is, she invites his attention by squatting. If not, she strolls off and he evaluates another ewe. Ewes utter low, tender, guttural vocalizations to their new lambs. It's a lovely sound, not to be missed.

High-pitched baaing. Frantic, high-pitched baas are stress indicators. They're the earmark of sheep in pain or very hungry and of individuals separated from dams, lambs, or flockmates. Medium-pitched baas are exchanged by adult sheep casually seeking nearby lambs, friends, or feed.

In Idaho's sagebrush steppe, a shepherd manages his flock on horseback. Proper handling is crucial; these sheep are part of a study at the U.S. Sheep Experiment Station to help native plant species compete against invasive species.

HANDLING SHEEP

The best way to handle sheep is quietly and gently. Move them no faster than their usual walking speed. Move the leaders, not individuals at the back of the pack. Decide what you want and how you can use the flock's natural tendencies to achieve your goal. Then enter the leaders' flight zones toward their rear until they set off in an orderly fashion; the rest will follow. Don't race about, shout, fling your arms, or otherwise frighten the sheep.

When approaching or driving sheep, don't gaze into their eyes. Wolves do that. So do dogs, cougars, and bears; staring is predatory behavior. Instead, focus on a sheep's nearest shoulder. She'll be far less likely to scuttle away.

Sweeten the pot. In a hobby-farm small-flock situation, it's worlds easier to lead sheep with a bucket of grain than to try to drive them. Treating friendly flock members with apple tidbits or crumbled commercial horse treats, or even scratching their shoulders or chests, will make them like you. Between dewormings, load the drenching gun with liquid molasses or honey and give willing sheep a squirt. They'll be easier to deworm next time. Take a tip from livestock handling guru Temple Grandin, who trained sheep to enter a scary tilt table and be briefly mobilized—for a grain reward. "Fourteen out of sixteen ewes," she writes, "returned for one or more additional passes." It's amazing what sheep will do for food!

Advice from the Farm

TIPS ON BEHAVIOR

Here are some helpful sheep behavior tips from shepherds who've seen it all.

Leaders, Followers, and Airheads

"There is a definite pecking order within all species. I have three different breeds of sheep, and they tend to graze within their breed. This could be because they were raised with their breedmates and are familiar with them and have their own pecking order.

"I was also told that different colors within breeds will stay together, and have found this to be generally true. As with any social species, there are leaders, followers, and airheads that are always in trouble, wandering off or being where they aren't supposed to be.

"As for intelligence, we have a few apple and pear trees in our yard. We occasionally let the ewes out to munch on the weeds. One ewe will stand in front of the tree and another will put her front feet on that ewe and start plucking fruit off of the tree. Seems pretty clever to me."

—Lynn Wilkins

Who's the Boss?

"Ewes definitely have a pecking order. Sometimes it's 'who's here longest,' 'who's biggest,' 'who's the most greedy,' 'why do I care,' 'I'm not feeling well, and

I'm not hungry,' 'who's the most aggressive,' 'I'm older, therefore I don't have time for these silly games,' and it depends on their breed, too.

"When I bring in a new ewe, she's immediately checked out by 'the bosses' and most often challenged to see where she will be in the pecking order. It also depends on where I had her for quarantine. I usually put newcomers in one of the horse stalls, away from the flock, to see how she responds to people and whether she's got a disease that I can see, and while she's there, I give her shots. Then I'll move her to a pen inside the barn with the flock around her. They'll come up and see her—at least most of them will; some just don't care—but they can't get to her because she's in that pen. If she's aggressive enough to 'butt' through the pen, I'll let her out in a day or two. If she's not aggressive, then I'll sometimes leave her in the pen until they figure she's just one of the flock and won't harass her at the hay feeder. Most of the time, she will just get turned out with the flock after being in the horse stall and establish herself.

"I've raised about ten to twelve different breeds, and while some breeds are categorized as 'aggressive,' not all of that particular breed may be. It depends on how they were raised too. A bottle-fed baby may not associate with sheep because she feels she is a 'people' instead of a sheep and just go on her merry way."

—Connie Wheeler

Mental Maps

"When sheep are used to a pasture and one gets out into the yard, it's relatively easy to herd them back through the gate (much easier than catching, certainly). I think they're uncomfortable in the strange area, though it looks the same (grassy and flat). But when sheep are brand new to a particular pasture, it's very difficult to herd them in—their mental maps do not even include an opening where you see and open gate."

—Lisa MacIver

Meeting Trouble Head-On

"The predominant ewes here seem to take a head-on stance and look directly into the challenger's eyes, and take a couple of steps toward the upstart. If the challenging ewe doesn't back off, the head-butting contest begins.

"Our ewes also use foot stomping quite often. Very occasionally with other ewes, but most usually when the herding dogs are in the pen. Sometimes with the guardian dogs—mostly, if they have lambs at side. Some seem to trust the guardian dogs more than others. Our Great Pyrenees is very respectful of Mom's wishes. She wants to check out her new wards, but if Mom stomps, she turns and walks away.

"And watch those bottle-raised cuties. Males can turn into monsters when they get older because they have no fear of humans. The worst working over I ever got was from a yearling wether we had a bottle raised. It was a nice September day and I was picking up in his pen. I bent over, and he took the opportunity to come at me from about 15 feet away. He had me down and hit me three times before I could get to a stick to crack him across the nose."

—Lyn Brown

John easily restrains one of our heftier sheep by raising his head and steadying him with a hand on his rump or flank.

Hire shearers who try not to gouge or nick your sheep and who never abuse them. Use a vet who likes sheep and handles them with care. Never snag sheep by their wool or ears, and avoid catching them by their legs if you possibly can. Grabbing or hoisting a sheep by its wool is very painful. Doing so can separate layers of skin and leaves nasty bruises.

Don't isolate a sheep unless you absolutely must. Even a penned ram needs a pal. Keep another sheep within sight to quell an ailing or injured sheep's fear. In their paper "Comfortable Quarters for Sheep in Research Institutions," researchers Viktor and Annie Reinhardt state, "Individuals show a multitude of endocrine, hema-tological, and biochemical alterations, stereotypic behaviors, and a marked increase in heart rate and respiration when isolated from other sheep." This study's recommendations translate well to small-scale sheep keeping.

It is best to make friends with your sheep so that when you need to do something with an individual, you can walk right up and slip a halter on her head. However, in the real world, that's not always easily done. Previously mishandled or wary sheep may never be hand tamed; even a gentle sheep, when injured or agitated, may elect not to be caught. Be patient. The old-timer may scoff at these measures, but compassionate handling pays off in contented sheep and fewer hassles.

Catch Me If Ewe Can

It's easier to snag a sheep out of a group than to chase down a frightened solitary member of the flock. Having other sheep present helps quell the target sheep's fear, and they'll block her escape route once she's captured.

Quietly herd the group into a corner. Draft an assistant to help, if possible. Because sheep have a blind spot of approximately 70 degrees directly behind them, you should try to approach your target sheep from the rear. Place one hand under her chin and raise her head, cupping the other hand around her rump; you can guide or restrain most sheep in this fashion. Or use a long shepherd's crook to gently snag the sheep around the neck, and then move in quickly to restrain her. Leg crooks work but can injure sheep; unless you know your sheep won't struggle, use a standard shepherd's crook instead.

Safety

On occasion, a panic-stricken, cornered sheep will jump high and hard when approached from the front. Be prepared to leap aside. Most adult sheep tip the scale somewhere between 75 and 400 pounds, and some have mighty horns. Frightened or angry sheep can badly damage or kill unwary humans. Never take your safety for granted.

Aggressive individuals generally warn before attacking and usually belong to one of two camps: either protective ewes with new lambs or, more likely, rams. If you breed sheep, rams are a necessity. But if you don't breed or you have small children or vulnerable adults in your family or neighborhood, please don't keep a ram. The verb "to ram" is derived from the Old English word ramm, meaning an intact male sheep. Apart from procreating, ramming is what unaltered sheep do best. Being charged by a ram is more than a bump on the backside. He may leap and crash into your spine or your chest. It'll bowl you over, and if you can't get back up, he might not quit. Always know where the ram is, even when visiting other flocks. Never, ever take any ram for granted!

Just Between Ewe and Me

It's a fact: once a sheep's feet are off the ground, she will remain stock-still. That's why sheep are 'tipped' for shearing and routine tasks such as hoof trimming and treating minor wounds. The chore seems daunting at first, but it's easy when you follow this routine.

Standing on the left side of the sheep, reach across and grasp her right rear flank (don't pull the wool!). Bend her head away from you, against her right shoulder. Lift her flank and pull back toward you. This puts the sheep off balance, and she will roll gently toward you onto the ground.

Hold the sheep on its side, then quickly grasp both front legs and set her up on her rump so she is slightly off center, resting on one hip. If she struggles, put one hand on her chest for support and inch backward until she's more comfy.

One caution: don't tip a sheep right after she has eaten; this position puts considerable strain on a full tummy.

CHAPTER FIVE

Health, Maladies, and Hooves

If you've been around sheep or shepherds very long, you've no doubt heard the old chestnut, "A sick sheep is a dead sheep," but that's far from true. What is fact is that sheep are exceedingly stoic creatures, and by the time anyone notices they're suffering, they may already have two hooves through death's door. The trick, then, is recognizing and addressing problems in their early stages. Following are some things to consider.

MAINTAINING A HEALTHY FLOCK

The best way to keep a flock healthy is to start with healthy sheep. Almost all illnesses surface after introducing new sheep to the flock. Even seemingly well sheep can be carriers. Thin, depressed, sniffling, coughing, or limping stock should be anathema. To keep out the serious problems such as scrapie, Ovine Progressive Pneumonia (OPP), and Johne's (YO-nees) disease, consider buying from certified-free flocks. If your flock eventually becomes certified, you'll have no other option.

Do your maintenance work. That means seeing that the sheep are dewormed, shorn, and vaccinated and that their hooves are kept in the pink. Sanitation is a must. Shelters, paddocks, and pastures should be tidy and free of poisonous plants, junk machinery, and all other debris sheep could get hurt on.

Feed your sheep quality hay, grain, and minerals, and keep things sheep friendly. Don't substitute grain mixes, minerals (including mineralized salt licks), or milk replacers that are not designed specifically for sheep. Finally, respect those rumens. Sheep can't digest feed until the right set of organisms becomes established in their

Dodger, our elderly Hampshire wether, dozes, smiling, in the springtime sun while visions of Cheerios and apple chunks dance in his head.

rumens. Switch feeds (even when going from hay to green grass) over a period of ten to fourteen days.

Know what well and sick sheep look like. Read the books and article resources listed in this chapter and in the Appendix. Know what you're looking for and know your sheep. Allow time to lean on the fence and watch them every day. Or haul a lawn chair into the paddock and just hang out. You'll enjoy yourself and learn a lot about flock social order. More important, when a given sheep behaves slightly out of round, you'll know there's something amiss, and you can investigate and treat him right away.

A sheep that doesn't eat with the rest or nibbles and then turns away is always suspect. If he spends his day lying down while the others graze, check him out. A depressed sheep standing hunched over, head down, is probably ill. Constantly runny or crusty noses and eyes, persistent sneezing or coughing, unexplained weight loss, limping, bizarre movements or head positioning, and intense rubbing on solid objects are definite red flags.

If you suspect there's something wrong that you can't easily fix, call a knowledgeable sheep person for advice. Better yet, phone your vet. Either way, so you don't forget something, take the sheep's tem-

Did Ewe Know?

Natives called sheep brought by Spaniards to the New World "cotton deer"; they were considered sacred and never eaten. The Tzotzil people of Southern Mexico still honor that tradition. Tzotzil shepherdesses constantly pray to John the Baptist, the Holy Shepherd, asking for protection and good health for their sheep. Their native Chiapas sheep are named, cared for, and respected as part of the family.

perature and write it down, along with his symptoms. Have the sick sheep penned so that examining him doesn't turn into a game of keep-away in the sheep pasture.

SHEEP AFFLICTIONS

Discussing all the maladies that befall sheep is beyond the scope of this book. Every shepherd (and especially those whose local vets are less than sheep savvy) should have a comprehensive, sheep-specific veterinary manual on the bookshelf. Alternatively, you can visit the University Resources section in our Appendix and download or print out bulletins relating to sheep health and sheep ailments. Keep these bulletins on hand for future reference. You won't be sorry you did. (See the Appendix for a list of some common sheep afflictions.)

YOU'RE CALLING THE SHOTS

Most shepherds learn to inoculate their own sheep. It's cost-effective, and veterinarians are generally pleased to make fewer farm calls. But you must consult with your vet to formulate a vaccination program specific to your area and your flock. Needs vary greatly from one locale to the next.

You can buy vaccines from your veterinarian, at many feed stores and farm stores, and by mail order from farm supply and biological warehouses. It's best to ask your veterinarian or an experienced sheep breeder to show you how and where to give injections, even if you routinely vaccinate other farm animals or horses. (See sidebar for tips.)

PARASITES

Like all other warm-blooded creatures (and some creatures that aren't), sheep are plagued by parasites. Some live inside the sheep, and some don't. Below are description of both kinds.

FLIES, LICE, AND TICKS

Sheep are annoyed by the same flying insects that annoy humans. Black flies nibble their ears, face flies buzz their noses, and stable flies make them stomp. But some of the worst sheep pests don't have wings at all: lice and the bugs we call sheep keds.

Sheep can become infested with sheep-biting lice and several species of blood-sucking lice: the sheep body louse, the African blue louse, and the foot louse. These pests are host specific, so unless you bring lice home on infested sheep, your flock will remain louse free. Louse bites cause intense irritation, so louse-infested sheep scrub their bodies on walls, fences, and anything else solid, in an effort to ease the itch. You'll know lousy sheep by their tattered, ragged fleeces; heavy infestations cause anemia and can sap

Did Ewe Know?

Humans can catch a number of diseases from sheep: sore mouth, ringworm, toxoplasmosis, Q-fever, campylobacter, salmonella, cryptosporidia, and tetanus. Contracting sore mouth and ringworm from sheep are fairly common occurrences. The others are not as common, but their consequences can be devastating.

Vaccinations

- Use disposable syringes and needles, and when you're through, dispose of them in a responsible manner.

- Use a clean, new syringe for every session. It's best to use a new needle for each animal. Sharp needles cause less pain and they work better. At roughly 30 cents per needle, it pays to stay sharp.

- Choose 16- or 18-gauge needles in 1/2-, 5/8-, or 3/4-inch lengths. Longer needles easily bend or break. Shorter ones are perfect for giving subcutaneous (injected under a pinch of skin) shots, and sheep vaccines are administered via that route.

- Swirl a vaccine bottle's contents to mix it. Don't shake; you want to avoid making bubbles.

- Pull back on the syringe's plunger, a little farther than the volume of the shot you'll be giving. While holding the bottle upside down, poke the needle through the rubber stopper. Depress the plunger to inject air and avoid the creation of a vacuum. Pull back a little farther than the dose requires, then gently press the excess back into the bottle, removing any bubbles you may have created.

- Always use a clean needle to withdraw vaccine from the bottle. A used needle contaminates the remaining contents. If you don't wish to use a new needle fore each sheep, insert a brand new needle into the bottle's rubber cap and leave it there. Arrange your syringe to it to withdraw vaccine (as

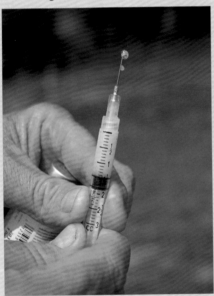

Specialized dose syringes such as this make it easy to administer liquid wormers and medicines, as well as fluid drenches.

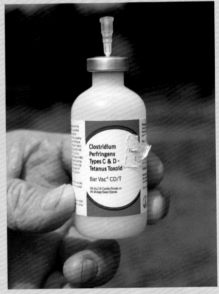

When vaccinating or injecting antibiotics such as this, it's always best to use a new disposable needle. Reusing the same needle is false economy.

above), then attach a new needle to the filled syringe. Voilà, you're ready to shoot.

- Part the fleece and give injections into clean, dry skin. Some vets recommend swabbing the area with alcohol; others don't.

- To give a subcutaneous injection, pinch up a fold of skin and slide the needle under it, parallel to the animal's body. Slowly depress the plunger, withdraw the needle, then rub the injection site to help distribute the vaccine. They can be given in the neck, over the ribs, and into the hairless area behind and below the armpit.

- Intramuscular shots are trickier, but you'll rarely have to give one except to administer antibiotic. Ask someone to restrain the sheep, then quickly but smoothly thrust the needle deep into the muscle. The preferred injection site of most vets is 5 or 6 inches below the withers (the highest spot on the sheep's back, at the point where its neck attaches to its shoulders). Always aspirate (pull back on the plunger about 1/4 inch) before you inject the contents. If blood sucks into the syringe, the needle pierced a vein. You must pull it out and try again.

- Store leftover vaccines and antibiotics in your refrigerator, following the instructions on their labels. Discard leftovers after their expiration dates pass.

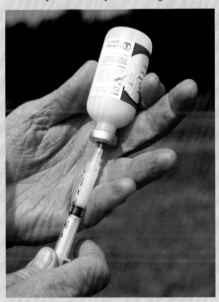

To prepare an injection, draw out a smidge more vaccine or antibiotic than you need, then depress the plunger to express air bubbles before administering.

John inserts a syringe into a sheep's cheek pouch, where paste substances such as wormer or probiotics are best deposited.

A wooly lamb scratches behind the ear. Scratching may be a sign that lice or ticks are present, so don't let behavioral clues such as this go unchecked.

Two winged pests wreak havoc on sheep: sheep-nose botflies and blowflies. Female nose botflies hover around sheep's faces and deposit maggots (larvae) in their nostrils. The maggots crawl up into their hosts' head sinuses and set up camp. When fully developed, they migrate back to the animals' noses and drop to the ground, where they pupate and eventually become adults. Adults are fuzzy, yellowish gray to brown, and a hair bigger than a honeybee. You'll know they're at work when your sheep stampede, stomp, snort, fling their heads, and press their noses to the ground. Sheep infested with maggots sneeze. Their breathing is labored, and blood-flecked snot may ooze from their nostrils. Fortunately, Ivermectin effectively kills botfly maggots when given as a 0.8 percent liquid oral drench.

Blowflies cause dreaded fly-strike by laying eggs in wounds and on inflamed skin (fleece rot) irritated by urine-soaked, filthy wool. Their eggs quickly hatch into maggots. These maggots secrete enzymes that liquefy their host's injured flesh. Fly-strike is incredibly nasty and painful; it can eventually lead to death from septicemia or toxemia. To prevent it, clean wounds as they happen. Clip wool away from the injury and apply antibacterial dressing or another wound treatment of choice, then smear insecticidal cream around the edges. Repeat daily until it heals. Eliminate fleece rot by keeping sheep tidy; continual moisture any place on a sheep's body irritates the skin and draws blowflies.

sheep's strength and immune systems. Infestations peak during the winter and early spring months, but they can be problematic all year. Sheep lice are fairly easy to kill: zap them with a spray, pour-on, dip, or lice dust product purchased from your vet or feed store.

Sheep keds are sometimes called sheep ticks, but they're blood-sucking wingless flies. They're flat, leathery, and reddish brown, and—at quarter-inch long—they're much easier to spot than lice. Sheep keds truck around their host sheep's bodies and poke their feeding apparatus through the skin to feed on blood and lymph. They cause itching and anemia just like lice; most products that kill lice dispatch sheep keds, too. The easiest time to treat sheep for lice and sheep keds is immediately after shearing. However, heavy infestations should be treated as they occur.

These sheep were part of a study by the USDA Agricultural Research Service into the sheep genome. Scientists searched for genes that affect resistance to disease, lamb survival, and fertility.

Shear dirty crotches and remove manure tags. Wash ewes' bloody backsides if they lamb during fly time. Clip wool away from rams' and wethers' penises. Don't give fly-strike a chance to occur!

WORMS

Sheep are troubled by four types of internal parasites: roundworms (nematodes), tapeworms, lungworms, liver flukes, and—only occasionally—meningeal (brain) worm. Only roundworms are common; all sheep have them to some degree. The trick is dispatching as many as you can, lest they damage or kill your sheep. Depending on their species, roundworms live in the stomach or small intestine. They are tiny; you're unlikely ever to see a roundworm. They feed on blood and body fluids and interfere with the digestion and absorption of foodstuffs. Badly infested sheep are thin and unthrifty; they usually cough. They're weak, sometimes anemic, and often suffer from diarrhea. Because they're run down, they're wide open to infectious disease. Lambs are more susceptible to roundworms than are adults.

In some parts of the country, lungworms cause problems for sheep that graze lowland and boggy pastures inhabited by snails. Sheep swallow the larvae while grazing; the larvae then migrate through tissue to their lungs. Affected sheep cough. Severe infestations cause fluid in the lungs.

Tapeworms, even substantial numbers of them, don't do adult sheep a lot of harm, but they can drastically affect the growth of lambs. Heavy tapeworm infestations, however, occasionally block a sheep's intestinal tract. Such an infestation can result in unnecessary death.

Liver flukes are another denizen of wet, low pastures. Like lungworms, their intermediary hosts are snails. Low to moderate infestations of liver flukes impact lamb growth; heavy infestations can kill adult sheep.

The meningeal (brain) worm affects all ruminants, wild and domestic; its natural host is white-tailed deer, but its larvae spend time in snails and slugs. When a sheep ingests an infected slug, the larvae travel up the spinal nerves to the spinal cord and the brain. This damages the central nervous system, resulting in a parade of bizarre symptoms and possibly death.

Internal parasitism is a complex topic. Your best approach is to study a wealth of research material, and then discuss the subject with your vet. When you do, take manure samples from several sheep. The vet can run fecal egg counts, assess your needs, and devise a deworming schedule custom-tailored for your flock.

HOOVES

Horse folks say, "No hoof, no horse." That principle applies to sheep, too: sheep with sore hooves are unthrifty. They're also unhappy.

TRIMMING HOOVES

A sheep with ragged, untrimmed hooves is an unhappy sheep. When hooves get long and roll under, they're uncomfortable and may cause a sheep to limp. When his feet hurt, a sheep won't graze as much or as often as his flockmates. He will lose weight. If a ewe is not providing the nourishment her in utero lambs require, they're born

John trims Dodger's hooves horse-style, with Dodger standing up. When you're as tall as John is, it can be hard on your back.

One of our miniature rams reclines in Premier's sheep chair, ready for John to trim his hooves. With the chair, chores such as hoof trimming and doctoring sheep are now a snap!

smaller than the other ewes' lambs. Untrimmed hooves can also contribute to serious sheep hoof disease such as scald and foot rot.

Trimming is probably the most important thing you can do for your sheep's hooves. Soil moisture and type, time of year, and breed influence how fast hooves grow. In general, plan on providing pedicures two or three times a year. Timing them to coincide with other labor-intensive procedures such as deworming, vaccinating, and shear-

ing makes good sense. However, avoid trimming hooves during high-stress periods such as extreme weather conditions, late pregnancy, or weaning. Hooves trim more easily when they're moist, whether from dew, rain, or snowmelt.

You'll need the proper tools. Most shepherds recommend hoof shears, as do folks who also keep goats. Shepherds accustomed to trimming horses' hooves often prefer horse-hoof nippers, a hoof knife, and a rasp. Shears or a hoof knife

With hoof and shears in hand, John is ready to begin trimming. The task is as simple as paring away the overgrown portions of a hoof. Irregular care could lead to curled ends, which cause sheep to limp in pain.

followed by a hoof plane work well, too. It's mostly a matter of taste, experience, and convenience: people tend to use familiar tools that they have on hand. If you're buying first-time tools, hoof shears are a best bet; they're inexpensive and easy to handle, and they make trimming a one-step job.

If your back is good or you have a helper, you can tip your sheep to trim their feet. Or one person can support the sheep while the other trims. In a solo operation, you can halter and tie your sheep and trim them standing up. You can squat beside the sheep, perch on an overturned bucket, or stand and lean over to trim.

Start trimming at the heel and work forward. Trim even with the frog (the soft, central portion of each toe), then trim the walls level to match. If the frog is especially ragged, you can touch it up with a knife, taking paper thin slices until you reach a hint of pink. The frog is a sensitive structure, so go no farther. When you're finished, the hoof should be flat on the bottom and parallel or nearly parallel to the coronary band.

When trimming a sheep with foot disease, trim the bad hoof or hooves last. Otherwise, you'll spread disease to healthy hooves. When you're finished, disinfect your tools to avoid infecting other sheep.

Foot Rot!

When shepherds spy a limping sheep, they tend to think the worst. But before concluding that your lame sheep has hoof disease, run a diagnostic evaluation. Watch the gimpy sheep from afar. Which foot or feet is she favoring? How badly is she limping? When you catch her, scan for foreign objects lodged in her toes or between them.

Advice from the Farm

WATCHING OUT FOR THE FLOCK'S HEALTH
Our experts talk about techniques and precautions in health care.

Giving a Trimming
"When trimming feet, notice the natural shape of the hoof. Trim down to the quick, making sure that each side of the hoof matches the other. Also cut off the point of the hoof to avoid those sharp hooves raking your arms and legs—or lambs, if your ewe paws her lambs to get them up."
—Connie Wheeler

Collecting a Sample
"Pinch the animal's nostrils shut for a minute. Have some sort of container under him so that you can collect the urine. He should pee right after you release his nose. Hang on, he is going to try to get away."
—Nancy Larsen

Splinting a Broken Leg
"I make a splint out of wood, vet wrap, and duct tape. I have never had to deal with a compound fracture, though some had to be set because the bone, at the break, was not meeting. All have survived and are doing well."
—Cathy Bridges

Knowing the Programs
"There are two scrapie programs: The Voluntary Scrapie Certification Program (tamperproof ear tags, tattoos, chips, five years to certify, and yearly inspection; tags cost around 75 cents each) [and] Scrapie Mandatory IDprogram (required for all sheep over eighteen months when they leave the farm of birth; free tags)."
—Connie Wheeler

Kneeling is this Scottish Blackface's way of telling you she's experiencing foot pain. Whether a signal of overgrown hooves or something more serious, such as foot rot, this calls for prompt investigation.

Next, examine the scent glands positioned between the toes and to the front of each hoof. Those glands get plugged. Gently squeeze the one on the gimpy foot. If a glob of waxy gunk pops out, that should do the trick. If nothing seems right so far, carefully trim all four hooves. As you do, watch for signs of disease. The ones you're most likely to encounter are foot rot and its little brother, foot scald.

Foot scald and foot rot are closely related. In fact, they share a causative agent: the bacterium *Fusobacterium necrophorum*. Although foot scald appears in the presence of *F. necrophorum* alone, the addition of *Bacteriodes nodosus* causes full-blown foot rot.

F. necrophorum is a common, hardy bacterium that dwells in soil and manure on virtually every farm on which livestock is kept. It causes thrush in horses and contributes to foot rot in cattle. It's an anaerobic organism (meaning that it can grow only in the absence of oxygen), so when animals are kept in dry, sanitary conditions, *F. necrophorum* poses no threat. But when hooves are continually immersed in warm mud and muck, bacteria invade the foot, often via a minor scratch or ding. The result is foot scald, a moist, raw infection of the tissue between the sufferer's toes. Foot scald usually affects one but not both front feet. It's nasty and painful, but its worst characteristic is that it often leads to foot rot.

Foot rot occurs when *F. necrophorum* is joined by *Bacteroides nodo-*

A pair of sheep face away from each other while grazing in green pasture. Before introducing new sheep into the flock, be sure they're wormed and vaccinated when they arrive and get boosters and rewormed in two weeks.

sus, another anaerobic bacterium that thrives only in the hooves of domestic sheep. It gains access via foot scald lesions and other injuries. If *F. necrophorum* is present, it sets up house in deeper layers of the skin, where it produces an enzyme that liquefies the tissue around it.

You can't miss foot rot. Affected sheep are very lame. Infected tissue is sleazy, slimy, and smelly. It's an odor you simply can't miss. Infection beneath the wall and sole of the hoof causes the horny walls to partially detach. More than one hoof may be involved.

Foot rot is treatable, but it's a long, costly, and time-intensive process—and, in most flocks, often not an entirely successful one. The key to foot rot control is to keep it away from your flock in the first place. If your sheep don't have it, they can't get it without coming in contact with *B. nodosus* bacteria through an infected sheep.

SAFETY WITH NEW SHEEP

If a prospective purchase limps, examine her carefully and think twice before you buy. When buying any sheep, you should play it safe by trimming her feet on arrival and then quarantining her, away from your flock, for at least three weeks.

Do the same with any returning sheep, whether they've been in a 4-H show, boarded at a veterinarian's facility, or away from your farm for any other reason.

As mentioned in previous chapters, you should not buy sheep at livestock sales. Many producers dump infected sheep at auctions. Even if the one you buy isn't infected, she has been exposed to infected sheep and held in pens where *B. nodosus* thrives. Foot rot bacteria grow slowly. Because seemingly disease-free sheep can harbor *B. nodosus*, you should consider any sheep from an infected flock verboten.

CHAPTER SIX

The Importance of Proper Breeding

Many sheep owners reach the point at which they wish to breed their sheep. (Lambs are cute little critters.) If you are thinking about breeding sheep, there are several things to consider and prepare for. You must determine what kind of sheep you want. You need to learn as much as you can about the breeding characteristics of your particular sheep and how to facilitate the process. You also need to know how to care for pregnant ewes and newborn lambs.

CHOOSING BREEDING STOCK

If you start with fifty-dollar sheep from the sale barn, you'll breed second-generation fifty-dollar sheep (if you're lucky). In essence, you get what you pay for. Buy the best breeding-stock sheep you can afford. If you want meaty ethnic holiday market lambs, choose sheep that breed out of season; pure or crossbred hair sheep would be a good choice. If handspinners' fleeces are your game, lay down the bucks necessary to own healthy sheep with the type of fleece you prefer. Pet breeders' sheep should be calm and manageable; berserk sheep give birth to spooky lambs. Always begin with sheep of the temperament, type, and quality you want to produce.

Buy breeding stock from shepherds who keep detailed records. Ask him or her what sort of lambs your prospective purchases have already sired or produced. Choose rams and ewes from multiple births that arrived early in the lambing season. Such ewes tend to produce like their prolific dams, and research indicates a ram's daughters' performance is strongly influenced by his dam's reproductive characteristics.

Consider buying a proven ram; ram lambs can be a gamble. An older, top quality ram may cost considerably less than a younger, mediocre one. The older guy can

Grazing alongside Mama, this cute lamb is a cross between wooled and hair sheep breeds.

shine in a hobby-farm setting, where he's comparatively cosseted and is required to settle fewer ewes. The same can be said for ewes. Large-scale producers routinely cull six- and seven-year-old ewes. With attention to their diets, some ewes are still lambing when they're twelve years old. Older ewes with outstanding production records are especially good buys for newbie shepherds; the old girls can teach them the ropes.

Did Ewe Know?

Ancient Chinese sheep were used to pull carts, carry burdens, and give children rides, especially in Xiongnu, where youngsters rode giant sheep as part of a hunting game. In Tibet and Nepal, thousands of sheep are still employed as pack animals.

Don't buy an ornery ram with one testicle. Avoid ewes with pendulous udders and balloon teats. Don't buy problems. Choose brood stock from scrapie-, OPP-, and Johne's-certified flocks if you can. Foot rot, lice, sheep keds, and excessive internal worm loads affect breeding performance. Sick, gimpy sheep are never a good buy.

If you have your heart set on breeding registered sheep, investigate the breed before you buy. Some carry genetic disorders you should be aware of (and avoid). Check registration papers before you buy. In most cases if the seller isn't the last recorded owner, he can't sign a transfer form, and you're stuck with expensive-grade sheep.

Keep your stock in the pink. Work with your vet to devise a tailor-made deworming and vaccination schedule and stick to it. Consult with your county agent and develop a sensible feeding plan. Trim those hooves. Shear in a timely fashion. Provide shelter from winter storms and summer sun. Content, healthy sheep deliver!

BREEDING

Most sheep are seasonal breeders. As autumn approaches (around August, in most locales), decreasing daylight triggers hormonal changes in ewes. They begin cycling (they come in heat) every sixteen or seventeen days until they become pregnant or breeding season ends (usually in January, when days start getting longer). While in heat, they "show" to the ram by squatting and urinating, and generally making themselves available; they're receptive for a day or two. Most ewes ovulate twenty-four to twenty-eight hours after a cycle begins.

Some sheep breed out of season, meaning they cycle all or mostly year-round. Wooly out-of-season breeders include Dorset, Merino, Rambouillet, Polypay, and Tunis. Most hair-sheep breeds cycle twelve months of the year. With especially good feed and care, these breeds can produce three lamb crops in two years.

Rams become infertile when their body temperatures soar too high. Steamy, sizzling weather, as well as stress, sickness, and overexertion from caroming around after ewes or battling other rams when it's hot outside, can do the trick. Shearing their scrotums and keeping rams indoors (sometimes with a fan trained on them) during especially hot days can help. Don't move or work breeding flocks during the heat of

This Scottish Blackface ewe seeks deep shade to escape searing summer heat. She finds shelter in a machinery storage structure.

At fifteen months, Abram is a full-grown Keyrrey-Shee ram. Rams as young as seven months old can begin breeding successfully; be sure to research and follow special feeding guidelines for young breeding stock.

the day, and make certain they can shelter in a ventilated, shady area when they want to. It pays to keep breeding rams cool. A mature ram in stellar condition can impregnate up to seventy-five ewes per season. A well-developed seven- to eight-month-old ram lamb can breed fifteen to twenty ewes, but he'll require supplemental feeding.

To be certain breedings occur, many shepherds fit their rams with marking harnesses. The crayon affixed to its chest assembly marks ewes when they're mounted by the ram. It's important to use only crayons designed for this purpose; others might not scour out of shorn fleeces. Change crayon colors every sixteen days, using the lightest colors first, to know exactly when each ewe settles.

Flushing is a practice endorsed by many shepherds; its purpose is to increase twinning (the USDA claims 18 to 24 percent more lambs are born to flocks where flushing is the norm). The process consists of feeding grain for seventeen days before exposing the ewes to a ram, then continuing for about thirty days longer. Grain is introduced slowly and tapered off when the flushing period ends. If you flush, don't overdo it. One-half to one pound of grain is plenty, depending on your ewes' weight and size.

Depending on the breed, ewes lamb 144 to 155 days after they're impregnated. Most shepherds figure five months from their last known breedings. Ewes shouldn't be overfed during their early pregnancies, when lamb growth is mini-

Meet Ewephemia, the baby daughter of Baasha, a double registered Keyrrey-Shee and Brecknock Hill Cheviot. Not yet a day old, winsome Ewephemia poses for the camera!

mal. However, because 70 percent of the lamb's growth occurs five to six weeks before birth, ewes—particularly ewes carrying twins and triplets (or more)—require supplementary feeding and careful monitoring. During these final weeks of pregnancy, ewes occasionally suffer rectal or vaginal prolapses and develop pregnancy toxemia or milk fever. Know what these conditions are and how to treat them; it just might save a ewe's life. Bred ewe lambs should be at least twelve to fourteen months old when they lamb. They need some cosseting, too. They should be kept in their own flock, away from bossy older ewes, and given a small amount of grain. It's best to breed ewe lambs to a smaller breed ram to cut down on problems at lambing time.

Crossbred lambs are often worth a bit less that purebred ones. However, it's better to produce an easily deliverable crossbreed lamb than a dead one—or a dead ewe—because the purebred lamb was too large for its immature mother to deliver. The practice of crossbreeding for this reason is very common in the livestock world. Cattlemen frequently breed heifers of the large exotic breeds to Angus bulls because Angus have small heads, thus producing calves that are more easily delivered.

THERE BE LAMBS!

As lambing time approaches, you'll want to prepare for the big events. You'll need to make many preparations before, during, and immediately after the births.

Build a Better Lambing Kit

Lambing is the most rewarding part of the shepherd's year. Usually the process goes without a hitch, but because glitches can occur, the shepherd should assemble a lambing kit to field possible emergencies. Here's how we do it on our farm.

We pack our lambing supplies in two containers. The one we take to the barn is a hard plastic step stool with a storage compartment inside. It's sturdy and tip resistant, holds a lot of gear, and—on cold, wet nights—sure beats sitting on the ground. In it we stow the following supplies:

- **Sharp scissors** to trim the umbilical cord to an inch or so in length. We disinfect them after each lambing and slip them in a plastic zipper bag (we use a lot of **plastic zipper bags** in our kit) to keep them clean.

- A **hemostat** to temporarily clamp on the cord if it continues bleeding (disinfected and kept with the scissors).

- **7 percent iodine** to dip the cord in after trimming. Some folks squirt iodine on the navel while the lamb is lying down, but it's much neater to keep **a shot glass** in the lambing kit to fill and use to dip the navel in while the lamb is standing.

- **Two flashlights**—we like packing a backup in case the first light malfunctions. We tuck them in another plastic zipper bag to keep them dry.

- **Lots of lubricant** for repositioning lambs. We like **Suberlube** and keep two squeeze containers in our kit.

- **Betadine Scrub** to swab a ewe's vulva before repositioning lambs.

- **Shoulder-length OB gloves**—sterile, individually packaged ones. They're harder to find than non-sterile gloves, but worth the search.

- A **sharp pocket knife**, so we don't have to use our umbilical cord scissors for routine cutting chores.

- A **digital thermometer**, the kind that beeps.

- A **bulb syringe** designed for human infants. It can't be beat for sucking mucus out of tiny nostrils.

- A **length of strong, soft rope** to pull lambs. This, the thermometer, knife, gloves, and other small items are stowed together in a single plastic zipper bag. We'd add a scourable paint stick designed for sheep or other IDmarker to the mix if we had more sheep, but our flock is so small that telling which lamb goes with which ewe isn't a problem.

- **Nutri-Drench** (the kind labeled for sheep) for weak lambs and exhausted ewes, including a **catheter-tip syringe** with which to give the stuff.

- An **adjustable halter and lead**—it's easier to move most ewes with one than without one. We use flat nylon alpaca halters for our girls.

- A **lamb sling**—quite a back saver!

- **Towels**—soft, old, cleanup toweling is stuffed in any remaining space.

Having once scrubbed spilled lube out of a foaling kit, we make certain the iodine, the lube, and the Nutri-Drench are individually double-bagged. Our other container is a lidded 5-gallon pail. It stays in the house, and its function is to keep our other lambing supplies centrally located. It houses the following:

- Ewe's milk replacer—we repackage it into plastic zipper bags and store 3 to 4 pounds in the lamb supplies container, the rest in a tightly sealed tin.
- A **plastic lamb bottle** with a **Pritchard teat** and several **spare teats**.
- A **flexible plastic feeding tube**, a **felt-tip marker**, and a **60 cc syringe** for tube feeding weak lambs.
- A **16-ounce measuring cup**, so we don't have to dig through the cupboards to find one when we need it; it's also great to store the spare

Pritchard teats in. It's a big one, so it can double as a milking pail, too.
- A **small whisk** for mixing milk replacer. All feeding supplies including the measuring cup are stored in a second plastic zipper bag.
- **Syringes and needles** go into another bag, along with an **elastrator and rings**.
- A **prolapse retainer and harness** resides at the bottom of the pail. We've never used it, but it's there if we need it.

A new Mama nurses her hungry offspring. Carefully shearing the area around your ewe's udder just prior to lambing time will help her babies find a teat.

Don't hesitate to ask for assistance, particularly from those who have had some experience with the blessed event.

Preparing for the Lambs' Arrival

About five or six weeks before lambing begins, administer booster vaccinations and dewormer. Immunized ewes transfer passive immunity to their lambs via colostrum, but those boosters must be current to count. CD/T (*Clostridium perfringes*, types C and D, and tetanus) is a must; your vet can tell you which others are necessary in your locale.

Four to six weeks before lambing, carefully, gently shear your ewes or crutch them. Crutching involves shearing only the wool surrounding a ewe's vulva and udder. With their warm fleeces gone, shorn ewes are more likely to lamb in a shelter where you want them instead of out in the snow behind a bush. After lambs arrive, shorn ewes seek shelter during foul weather instead of letting their little ones chill. Finally, shearing or crutching off filthy wool tags makes it easier for new lambs to find a teat.

At the same time, begin supplementing their diets with grain. Start

In her final labor stage, our ewe Rebaa has sequestered herself in our sheepfold. Early that morning she began circling, pawing, and lying down and getting up again—all signs that her lambs were on their way.

with 1/4 pound per ewe per day, gradually increasing rations until they're up to between 1/2 and 2 pounds apiece, depending on their size and body condition. If fat, bossy ewes eat more than their share, separate them into smaller groups.

If the hay your ewes eat or the pasture they graze is selenium deficient, they'll need selenium/vitamin E or Bo-Se shots four weeks before lambing. Ask your veterinarian about these; he'll know about deficiencies in your region. Also, watch ewes closely for the last two weeks before lambing. Should they succumb to milk fever or pregnancy disease, it's likely to happen then.

Assemble a lambing kit or update the one you already have. Build or set up jugs (individual post-lambing pens for each ewe and her lambs) in a well-ventilated, draft-free area in a shed or barn. On hobby-farm scale, allow one for each five ewes. Build 4'x4' jugs for small breeds, 5'x5' jugs for larger ones. Use wood or welded hog panels to construct them; if drafts might be a problem, opt for solid wooden walls. Fit each jug with a hay feeder and water bucket, and bed it with bright, fresh, long-stemmed wheat straw or other suitable bedding (not sawdust). Don't use five-gallon food-service buckets for water; lambs have been known to drown in them.

ARRIVAL

Now keep an eye on those ewes! As lambing day approaches, their udders enlarge (sometimes more than seems possible); by lambing day, teats appear round and firm. Vulvas get floppier and take on a rosy hue. As lambs drop into position, most ewes become slightly swaybacked, and hollows appear in front of hipbones. Just before lambing, many hie themselves off to a far corner of the pasture or seek a dark, cozy corner in the barn. Don't pen them up unless you must; ewes like privacy and want to choose their own birthing spots. As labor begins, ewes paw as though nesting, throw themselves down, get up and circle, then lie down again. Within about four hours, barring complications, they'll deliver their lamb or lambs.

When the first lamb is born, the mom begins murmuring to the new babe. She'll lick and lick the lamb. Don't wipe the newborn dry! Licking is part of the bonding process. However, as soon as the lamb appears, carefully strip fluid from the nostrils, then get back out of the way. Don't imprint newborn lambs; their mother might decide they're yours and let you raise them! Wait to see if another lamb is coming. If one is, the ewe will dig, circle, and flop down again, leaving her first lamb to deliver the second. Creep forward and retrieve the first lamb. Use sharp scissors to trim the umbilical cord about 1 inch from the tummy, then dip the cord in, spray, or soak it with 7 percent iodine. *Don't omit this step.*

When all of her lambs have arrived, check the ewe's udder. Strong wax plugs in teats sometimes prevent lambs from nursing. Make certain milky fluid squirts out of each teat, then watch until the lambs begin suckling. Next, move the little family to a lambing jug. Carry the lambs at ground level where

One of Rebaa's newborn lambs, Ewelanda, takes a rest after her first feeding. Little brother Wooliam feeds while Ewelanda relaxes, her tummy filled with warm colostrum.

Advice from the Farm

LAMBS AND LAMBING

Advice about lambing from the experts.

Turning to the Jug

"Jugs keep a ewe confined to where her lambs have a chance to get a drink, and it forces the ewe to realize she has two or more lambs. Some ewes won't accept both lambs and have to be tied for a day or so. Every shepherd should have one or two jugs, even if they pasture lamb. I don't feel any jug should be larger than 3 by 4 feet."

—Barbara Burrows

Lifting a Finger to Assist

"Lambs are rarely so far in that you can't reach them with your hands, and even better if you can just put your finger in to pop out a stuck leg or something. In the seven years that I've been here, I've only assisted once (twins, each with stuck front legs; just popped them out with a finger) and that same ewe had a VERY mild prolapse, which went right back in after lambing. I used a ewe spoon to keep it in while she was carrying."

—Laurie Andreacci

Practicing with the Puller

"To practice, put a stuffed animal inside a paper bag and, without looking, reach in and figure out what is head and front legs. With the lamb puller, you loop the string over your fingers and reach in, so you'd have to do it all with one hand to wiggle the noose off of your hand and onto the lamb—you probably won't need to do it, ever—but if you practice with the paper bag you at least have an idea of how it works. Keep in mind that there isn't that much 'free space' inside a ewe, and you'd have insides working against you."

—Laurie Andreacci

Knowing if They're Eating

"If they're not slab-sided and aren't sunken in, they're usually fine. Mom most generally won't let them nurse for very long at at time, unless her udder is really full of mlk. If their mouths are warm (a sign that they're healthy; a cold mouth is a sign of hypothermia), feel their tummies. Are they full? If they are, the lambs are eating. I wouldn't bottle feed them unless they start having problems or it becomes very obvious they're not geting enough milk."

—Kim McCary

Slinging to Avoid Flying Lambs

"One thing Ihave in my lambing kit is the lamb sling. Since we usually have twins and often triplets, one person can gather up the lambs in slings and hold them in front of the ewe, and she will follow along into the jugging pen with little or no hesitation. If you pick up the lamb, the ewe gets confused; ewes do not understand the concept of 'up.' Nature did not equip them to look for flying lambs."

—Lyn Brown

their mom can see them. If she's halter trained, halter and lead her while someone else carries the lambs. It'll save a whole world of hoo-ha!

It doesn't always work like this. While most lambings proceed without incident, sometimes things go terribly awry. Lambs are stuck or malpositioned; ewes die or reject their lambs; lambs are born too weak to stand and suck. You must be prepared for every eventuality.

You must understand what a normal delivery looks like so you can spot problems early. If a ewe needs help—if her lamb or lambs are positioned incorrectly or if for any other reason she can't deliver them—you'll need to call the vet (pronto) or assist her yourself. Buy a book on lambing (you'll find some listed in the Appendix) or print out one of the excellent free lambing guides available online from various state extension services (also listed in the Appendix). Memorize the instructions or stow them in your lambing kit. Slip inkjet copies into plastic sheet protectors and seal them with tape; don't lose those instructions to dampness when they're most needed.

Be there when lambs are born. Monitor them periodically throughout the day and roll out of bed to check your ewes at intervals throughout the night—or set up temporary camp in an out-of-the-way corner of the barn. Keep checkups low-key, but check often. Every few hours is sufficient.

Though no one could resist a face like this, new lambs need more than just TLC. Proper nutrition, disease prevention, and warm shelter are at the top of a long checklist.

Bottle Feeding

BOTTLE FEEDING ORPHAN LAMBS WITH LYN BROWN

"What I usually do to get my new lambs started is sit with my legs crossed, tuck the lamb in the middle in a sitting position (front legs straight and butt on ground). I cup my left hand under the lamb's jaw and open the mouth and insert the nipple with the right hand; once the nipple is in the mouth, I balance and steady the nipple with the left hand that is still under the jaw.

"In other words, I keep the bottle and nose aligned so that the lamb doesn't spit or move the nipple to the side or back of the mouth. I elevate the bottle with the right hand only enough to avoid the lamb sucking air. In this position, you can feel the lamb's throat with the heel of your hand, and you know if it is swallowing.

"If you elevate the bottle too much, the milk can pour into the mouth, and if the lamb were not swallowing, the milk could enter the lungs. I try to keep the bottle as level as possible while keeping milk in the bottle cap and nipple. Of course, that means the more the bottle empties, the more tilt there needs to be.

"Most people kill their first bottle baby with kindness; they overfeed it because the lamb cries and they think it must be hungry. I know I did. I follow this feeding schedule strictly (no exception.). If our lambs cry between feeds, we feed them Pedialyte or Gatorade. That won't hurt them as far as enterotoxemia goes and gives them electrolytes while filling the void for them.

- **Days 1–2:** 2–3 oz 6x/day (colostrum or formula with colostrum replacer powder)
- **Days 3–4:** 3–5 oz 6x/day (gradually changing over to lamb milk replacer)
- **Days 5–14:** 4–6 oz 4x/day
- **Days 15–21:** 6–8 oz 4x/day
- **Days 22–35:** work up gradually to 16 oz 3x/day
- **At about 6 weeks**, I begin slowly decreasing the morning and evening feedings and leave the middle feeding 16 oz., until I eliminate the morning and evening bottle entirely (remember, they are eating their share of hay or pasture by now). I continue with the one 16-oz bottle for about two weeks, then eliminate the bottle feedings entirely.

"By making changes gradually, you can observe changes in the condition of the animal and judge and adjust accordingly. Gradual changes also avoid the complications (some of which can be fatal) of sudden changes in diet. Whatever you do, when you buy milk replacer, use lamb replacer. All-purpose milk replacers and calf replacers do not work well with lambs."

—Lyn Brown

Advice from the Farm

BANDING TAILS

Our experts discuss banding protocols.

A Different View

"I don't band the tail on the first day, as when I've done it in the past, the mom smells the alcohol on the band and then she won't lick the lamb's bottom. When lambs first poo it's not solid, but very soft, and if mom doesn't lick it away, it sticks. That's what happened with some lambs last winter. The mom never did lick her lambs' butts, so I was there to wipe away anything that came out (I won't be doing that again!)."

—Laurie Andreacci

Band the First day

"We use the elastrator band, and we band the first day. It has been our impression (no scientific evidence, just observation) that newborns are equipped with pain-blocking mechanism for the birth process, and if we do the tails that first day, they don't seem to feel it as much as if you wait a day or two.

"Either way, it is only painful until the tail goes numb, which in our experience is only a half hour or so. We've had no complications or infections from the banding process. that's with from forty to a hundred lambs per year.

"If you decide that you are going to leave tails on, which is perfectly alright, then you must be observant about sheep developing loose bowel movements. The tail can stick to the body, blocking off the anus and putting the animal in serious medical jeopardy. You even need to watch for this on the young ones, before the tails fall off. Make sure you keep their little bottoms clean."

—Lyn Brown

Don't Band a Weak Lamb

"I agree with Lyn about it being less invasive the first few hours of life. Even on day two or three, there seems to be a difference in the reaction in the lamb vs. the first twenty-four hours. I absolutely will not band a weak lamb, though, and I am sure that Lyn will concur with that."

—Cathy Bridges

New Lambs

Not everything is clover once lambs are born. Lambs are susceptible to diseases and conditions as diverse as constipation and scours (diarrhea), pneumonia, acidosis, enterotoxemia, tetanus, polio, and white muscle disease. Learn all you can about these problems before your lambs arrive; the resources in our Appendix will point the way. Lambs are delicate, wee creatures and may need your help to survive.

Weak lambs may need to be tube fed until they're strong enough to stand and suckle. Passing a feeding tube is a daunting task to most first-time shepherds. Ask your vet or a sheep-savvy friend to show you how in advance.

Orphan and rejected lambs as well as weak lambs with a suckle reflex can be bottle fed. Every shepherd should keep sheep's milk replacer and feeding bottles on hand. Because newborn lambs must ingest their mother's colostrum—first milk liberally laced with antibodies and produced by the ewe—for roughly twenty-four hours, hand milking is another skill best learned in advance.

New lambs must be kept reasonably warm. Some shepherds use heat lamps above wintertime lambing jugs, but this can be risky indeed. If you use them, make certain they're fastened very securely (don't hang them by their cords) and far enough from flamma-

Newly banded Wooliam explores his new surroundings. Some experts say that lambs have a mechanism that blocks pain before, during, and shortly after birth.

ble bedding that they won't trigger a fire. Blankets are an option but not usually a good one; blanketing newborns masks their scent, and sniffing is how ewes recognize their lambs. Lambing jugs with solid side panels, set in a draft-free section of the barn and deeply bedded with hay or long-stemmed straw are sufficiently warm in most cases.

TAILS

Lambs are born with long tails; tradition demands that we whack them off. But there is method to this madness. Long tails become urine soaked and encrusted with manure; the mess attracts blowflies. Blowflies lay eggs in the filthy mass, and hatchling maggots create fly-strike. Not a pretty sight.

Lambs are usually docked (their tails removed) when they're a day or two old. Although some shepherds merely lop them off, most use Elastrator bands—thick, strong rubber bands that cut off circulation to the tail, causing it to slough off in a week or so. It's considered the most humane method.

Until recently, tails were docked close to sheep's bodies. However, close-docked ewes often prolapsed before or while giving birth. Nowadays, tails are left longer, often long enough to cover ewes' vulvas. A good way to determine where to apply a band is to lift the lamb's tail and band it where the web of skin beneath the tail meets wool.

Banding is probably more than a little uncomfortable; most lambs race around for ten minutes or so until their tails go numb. But after that, they're fine. It may not be fun for them, but it certainly beats fly-strike.

The adventurous Wooliam explores his new world, flaunting his cute, au natural tail. Because soggy and manure-laden tails attract egg-laying blowflies, it's always wise to dock them.

Lamb's soft wool and pint size invites cuddling, especially from children. Handling newborns should be kept to a minimum, however, since scent is a ewe's primary means of identifying her offspring.

CHAPTER SEVEN

Fleece: Shearing, Selling, Spinning

Unless your sheep are hair sheep, they absolutely must be shorn. This is necessary not just to harvest fleece, although that's a valid reason; it's the healthy and humane thing to do for several reasons. For one thing, most unshorn sheep keep growing wool. For another, rain-logged fleece pulls on tender skin, causing skin separation, pain, and sores. Masses of filthy fleece invite fly-strike. Megawoolly, unshorn sheep are prone to heat stroke. Pregnant ewes may lie down, roll onto their backs, get stuck, and die.

Whether made from the fine wool of a Merino or the felt from scraps obtained during your first attempt at shearing, you can make a wide variety of garments. You can also pick up some hefty bucks from others who wish to use your fleece, such as handspinners.

THE FLEECE COMES OFF

That the wool has to come off is a given, but shearing is a stressful time for sheep (and shepherd, too). Make the process as neat and painless as you possibly can. First, learn which time of the year is best for shearing your sheep. Next, decide whether to hire a professional shearer or to learn to do the job yourself.

TIME IT RIGHT

Although mild-climate wool producers sometimes opt for winter shearing, sheep are traditionally shorn from March through May. Some breeds are shorn twice a year, in spring and again in early fall. Your shearing schedule will depend on a number of factors, including the type of sheep you have and your locale.

Before shearing: Baasha models a year's growth of sun-bleached fleece.

Many shepherds shear their ewes before lambing. Otherwise, new lambs seeking their mothers' teats sometimes find and suck wool tags instead, missing out on vital early meals of colostrum. Shorn ewes take shelter when it's cold, whereas sheep with warm, heavy wool may stay outdoors where their tender, newborn lambs are apt to freeze. The downside to pre-lamb shearing is that late-term ewes must be handled more carefully than other sheep. It's wise to have them shorn well before lambing.

Unless you can temporarily place them in a warm barn or blanket them, don't shear sheep until the last winter storms have passed. It takes most sheep six weeks to grow an insulating layer of wool, and they're especially vulnerable for several days post shearing. By the same token, in steamy southern locales, you should shear before warm weather sets in. Heavily fleeced sheep, especially dark-colored individuals, sizzle under a sultry summer sun.

THE PROFESSIONAL SHEARER

If you can hire a competent shearer, you're wise to do so. The job requires know-how, a strong back, and specialized tools. Shearers are in very short supply, and good ones are scarcer still. If you can hire a shearer, choose one with a sterling reputation. The shearer should handle sheep humanely and remove their fleeces in readily market-

Did Ewe Know?

Wild sheep have hair rather than fleece; they do grow a woolly winter undercoat, which they shed each spring. However, a 6,000-year-old figure of a woolly sheep was unearthed at an archaeological dig in Sarab, Iran, proving that wool sheep had been developed by that early era. By the ninth century BC, the Greek poet Homer praised the whiteness and quality of wool produced in Thessalia, Arcadia, and Ithaca.

After shearing: The beautiful Baasha appears much darker with her sleek new 'do.

able condition. To locate shearers, ask other shepherds for recommendations, scout for business cards and flyers on the bulletin boards of feed stores and large-animal veterinary practices, call your county agricultural agent or state vet school, and peruse online livestock directories.

Skills and Fees: Ask for and check shearers' references, and explain your expectations up front. Shearers are paid by the head, so they want to work quickly and be on their way; some aren't willing to slow down to bow to humane demands. If that's the case, look elsewhere or learn to shear your own sheep.

The epidermal layer of sheep skin is thin and tender, so even the best shearers sometimes nick, scrape, and ding your sheep. But multiple superficial wounds, deep cuts, and teat and penis injuries are unacceptable, as is hauling sheep around by their wool. A fleece should be removed in a single piece.

Reshearing a given area results in short bits called second cuts. They drastically reduce market value and are the mark of a careless or inexperienced shearer.

Shearers charge by the head, usually in the three- to fifteen-dollar range. They figure fees based on how far they must travel to your farm, the number of sheep they'll process, and the type of wool they'll be handling. If you have a small flock, be

Did Ewe Know?

Today's method of shearing sheep—the Bowen Technique—was pioneered by New Zealander Godfrey Bowen in the 1940s. His brother, Ivan Bowen, won the first International Golden Shears title in 1960, along with four more world championships. At eighty-four, Ivan still competes in Golden Shears veteran competitions, shearing a sheep in about sixty seconds. He does a hundred push-ups each morning to stay in shape.

Better Fleece

PRODUCING BETTER FLEECES

Whether or not you jacket your sheep, there are ways to produce better fleeces.

- Clean up your pastures. Weeds, burdock burrs, thistle fluff, and seed heads all work deeply into fleece.

- Avoid hay and straw put up with polypropylene strings. Short bits of twine sucked into the baler end up tangled in fleeces, and it's the hardest contaminant to remove. Keep an eye out for stray plastic feed-sack fibers, too.

- Don't use feeders that allow sheep to bury their heads in hay. Standard big-bale feeders are especially poor choices.

- Don't toss hay at feeders when there are sheep in the path. When feeding concentrates, try to avoid pouring grain onto the heads of sheep jostling for position at the feeder.

- Choose contaminant-free, long-stemmed straw for bedding; reject straw full of weeds, seeds and chaff. Waste hay will work if it's free of dust, mold, and fleece contaminants. Don't use sawdust or wood shavings unless they're tipped by straw or hay.

- If you use a marking harness on your ram or you spray paint numbers on ewes and lambs for identification purposes, choose chalk, crayons, or paint that easily scours out of fleeces.

- Never allow your sheep to be shorn when they're wet. Damp fleece yellows and molds.

- Keep your sheep dewormed and free of sheep keds, lice, and other skin irritants.

- Follow a well-balanced feeding program. Well-fed sheep grow twice as much wool as scrawny, sickly sheep.

- Vaccinate for diseases prevalent in your locale. Sick and otherwise stressed sheep suffer wool break when their fleece growth is temporarily interrupted and then resumes growing, leaving a weak spot in each wool fiber.

- Choose a breed that produces the type of wool you want to market. If you're not sure, scope out the resources at the end of this chapter or join a fiber arts e-group.

- Consider black sheep. Most handspinners prefer black (which actually ranges from ebony to shades of pale gray) or uniquely colored fleeces such as those grown by Icelandics, Shetlands, and Navajo-Churros. Quality white fleece sells too, but colors are all the rage.

Positioning sheep on a grooming stand, as shown here, is helpful for new hobby farmers who may find professionals' moves cumbersome. The elevation allows the shearer to move around the animal for a precise cut rather than awkwardly bending over a fidgeting sheep.

mercial wool prices are too low to make this economically feasible, although some will charge less if they take your wool.

Shearing Prep: Once you've scheduled the services of a recommended shearer, start preparing for his or her arrival. The night before the visit, pen your sheep in a clean, dry, roomy section of a barn or shed. Sheep can't be shorn if they're wet—and that includes dew or frost.

Sheep shouldn't eat for eight to twelve hours before shearing. The positions the shearer places them in are fairly comfortable for sheep as long as their stomachs aren't packed. Uncomfortable sheep are likely to struggle, and struggling sheep often get nicked.

Provide a clean, level shearing surface. Have it in place before the shearer is scheduled to arrive. Two possibilities are a clean tarp or two sheets of plywood with the space between them sealed with duct tape. Shake the tarp or sweep the wooden surface between sheep. The shearing floor should be situated in a well-lit, well-ventilated, covered area. Have heavy-duty extension cords on hand if your shearer needs them (ask when you make your appointment); flimsy household cords won't do.

Arrange for additional help if you need it (and you probably will). The shearer's basic fee doesn't include snagging reluctant sheep from the flock or hauling them to a pen when finished. Stock up on season-appropriate hot or cold drinks for your shearer and helpers; wrestling sheep

willing to pay premium per capita, mileage, or setup fees to obtain quality service. It's not cost effective for professional shearers to drive many miles to shear a flock of fifty head or fewer.

Another option is to check with other small flock keepers in your locale and arrange everyone's shearing for the same day or weekend. If you do, ask the shearer to visit each farm; don't gather sheep at a central location. Commingling sheep easily spreads disease from flock to flock.

In days past, many shearers accepted shorn fleece in lieu of fees. Today's com-

Abram is haltered and ready for shearing. A drop cloth or plastic tarp such as this protects shorn fleece from contaminants. It should be swept or shaken clean between sheep.

is thirsty work. If shearing overlaps lunch or supper time, provide sandwiches but not a heavy meal.

Just Between Ewe and Me

Whether or not you jacket your sheep, there are ways to produce better fleeces.

- Wear old shoes. Sooner or later, someone will pee on your foot.
- Don't wear shorts. Scrabbling sheep hooves hurt—a lot.
- If you thing the gentlest sheep will be the easiest to shear, you are mistaken.
- Newbies don't peel off nice, neat one-piece fleeces—ever.
- After the sheep are shorn, you can touch up your motley job with horse clippers. But clean them between each sheep!
- Mistakes grow out, and sheep forgive you—no matter how funny they look.

Pen your nonworking dogs, and keep herding dogs away from the shearing floor. Dogs worry many sheep, causing undue stress and fidgeting. Stay with your flock until the job is finished. Make certain it's done to your satisfaction.

Keep antiseptic at the ready to treat nicks and dings. The shearer shouldn't have to wait while you fetch it. If you trim hooves and deworm on shearing day, do it after the sheep are shorn. Don't expect the shearer to hold each sheep while you trim or treat it.

Advise the shearer of special needs before shearing begins. Elderly, chronically lame, and recently ill sheep require gentler handling, and a lone wether in a flock of ewes could have his penis zipped off if a speedy shearer doesn't know it's there. If it's cold and you haven't proper shelter for shorn sheep, ask the shearer to

A strange dog's approach puts our wee flock on alert; even the lambs are wary. Since sheep remain constantly aware of their surroundings, make sure your shearing area is secured from dogs and other animals whose presence may stress your sheep.

use winter combs that leave a short layer of fleece on your animals. When the shearing is done, pay up. Don't expect the shearer to mail you a bill.

WHEN THE SHEARER IS YOU

Amateurs can certainly shear their own sheep, but don't expect a professional-looking job the first few times. Invest in quality tools. Determine which sheering position—sitting or standing—will work for you. Your back and your sheep will be grateful you did.

Handheld Tools: When you have only a few sheep to shear, or if money is a concern, traditional hand shears will do the trick. A high-quality, triple-ground shear with 6.5-inch blades costs about twenty-five dollars, and a second set of twenty-dollar minishears makes touchups a snap. Add a six-dollar sharpener to refresh their edges between professional grindings, and you're in business!

Keep your shears sharp. Hold them parallel to the animal's body. While you're learning, work slowly and cautiously, or you're likely to cut your sheep. As you gain experience, you can work considerably faster.

If you're short on time, have a lot of sheep, or cost isn't terribly important, invest in good electric shears. Shears are not horse or dog clippers; they are heavy-duty machines equipped with heads specifically designed to shear sheep. Oster, Andis/Heiniger, Lister, and Premier build quality shearing machines. You'll find a comparison of handheld electric shears in Premier's free sheep and goat supply catalog (see Appendix).

Plan to spend 250 to 350 dollars for electric shears and roughly forty to fifty dollars for each comb and cutter set. You'll need twice as many cutters (extras run about five dollars each), as they dull much faster than combs. Pros shear up to twenty sheep per comb and ten per cutter; novices should allow two cutters and a comb for each pair of sheep. Combs and cutters can be professionally resharpened about ten times; don't try to resharpen them yourself! You'll also need clipper oil; choose a clear product to avoid staining fleece.

While in use, cutters with clipper oil at three- to five-minute intervals. Clean and coat blades with clipper oil within an hour after you finish shearing; store them in plastic clipper-blade boxes. Stow everything in a sturdy case or padded bag. Take care of your shearing gear to make it last.

Taking a Seat or Standing: Professional shearers follow a set shearing protocol, swiftly setting and swiveling sheep in a series of positions whereby they're immobilized, and marketable fleece is zipped off in a single piece. Hobby farmers can learn this set of moves or shear their sheep while standing. Weekend shearers of limited strength and battered back will likely prefer the latter ploy.

A shearing stand makes the job a whole lot easier. A goat-milking platform or the sort of stand designed for beautifying show lambs works very well. If you don't have one, not to fret; you can shear sheep standing on the ground.

Secure the sheep's head. Show sheep stands are fitted with a cradle apparatus that does just that; a fence-mounted version is available from Premier. If you have neither, use a halter and lead to tie the sheep to a sturdy fence. Use a slip knot in case it gets tangled and needs to be set free fast.

John shears Abram the old-fashioned way, using traditional hand shears. This is often the best method for small flock owners (though Abram isn't certain he thinks it's a good idea).

Dodger sports a sheep halter and lead—just the right gear for shearing time. Hobby shepherds have many varieties of halters to choose from, including popular show halters such as this, one-size-fits-all plastic rope halters that cost under three dollars each, and flat nylon halters designed for miniature horse foals and yearlings.

Be patient. Most sheep aren't thrilled with this situation. Start at her head and work carefully to remove desirable fleece in a single piece. When it's off, shear her face, lower legs, and as much belly wool as you can, then set her on her butt to finish the rest. Take long, bold strokes with electric shears, and keep the comb flat against your sheep's skin. If the comb rises, don't make a second sweep—at least not until the fleece has been removed.

Don't expect instant success. Your first shearing will probably ruin the fleece for commercial purposes and embarrass both you and your sheep. But take heart; you can use the wool for home felting projects or quilt batts, and the more you shear your sheep, the better you'll get.

Whether a professional shears your sheep or you do it yourself, you must take proper care of the shorn fleece if you plan to sell it. This includes caring for the fleece while it's on the sheep as well as after it's removed. Some ingenious ideas have been developed for such purposes.

SHEEP CHIC

In 2001, Australian Wool Innovation Ltd. commissioned a study to determine the economics of jacketing sheep. Twelve hundred Australian sheep in six flocks (including a control group) took part in the study. Its findings: covers significantly improved wool yield, reduced fleece contamination, and drastically reduced fleece rot and fly-strike. They didn't affect fleece weight, micron count, or staple length and strength. The paper recommends that sheep coats be constructed of tightly woven ripstop nylon treated for UV resistance, with expandable fronts, rump coverings, and roomy leg straps. The experiment was a rousing success.

Did Ewe Know?

A single, magnified wool fiber resembles a stack of flowerpots piled one inside another, all facing outward from the base. Stroked through fingertips from base to tip, its shaft feels smooth; tip to base—stickery. This scaly outer covering is called the cuticle; it causes fibers to interlock with one another. Inside each fiber, a cablelike cortex imparts strength. A wool fiber is so springy and elastic that it can be bent more than 30,000 times without harm. Stretched, considerable crimp allows it to spring back to shape.

Most American small-scale producers of handspinners' fleeces jacket their sheep. Besides keeping fleeces clean, covers keep sheep warmer after shearing and cooler in the hot summer months (white jackets on colored sheep are especially useful). Their failings: shepherds spend considerable time and effort making, laundering, repairing, changing, and making certain sheep coats fit. Some felting of fleece may occur in some breeds, especially along the spine and around necklines. Considering that coated fleece routinely sells for considerably more than the price of unjacketed fleeces, most shepherds consider it a pretty fair trade-off.

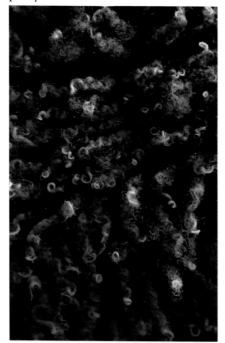

Whether light or dark in color, wool samples such as this are evaluated according to set guidelines for grade, class, and quality.

It's fairly easy to sew your own sheep coats, but you can buy them, too. You'll need several sizes per sheep to allow for fleece growth. However, the size one sheep outgrows may neatly fit a flockmate—at least for a while. Ready-made sheep coats run ten to eighteen dollars each, depending on size.

You have some choice in materials. If coats aren't crafted of breathable fabric, the wool they cover can mildew or mold. Uncoated nylon and woven plastic fabrics work well, but canvas is usually a poor choice, especially in damp climates.

Most commercial models slip over a sheep's head, then its legs are threaded through the leg straps. It's fairly easy to jacket sheep that are accustomed to wearing covers. Things can get terribly exciting, however, the first time your flock gets dressed.

SELLING THE FLEECE

Nowadays it usually costs more to have sheep shorn than their wool is worth. Yet a growing legion of small-scale sheep keepers—hobby-farm shepherds, if you will—show a tidy profit (or at least pay for their flocks' annual upkeeps) by producing quality fleece for handspinners. There are other markets and uses for the fleece as well.

THE LINGO

To deal in wool, you must learn the lingo. First, familiarize yourself with terms that refer to the fiber itself. The value of handspinners' fleece is deter-

Advice from the Farm

SHEEP COATS AND WOOL SELECTION
Our sheep experts talk about sheep coats and wool selection.

Nylon Coats

"I've seen sheep coats made of canvas and have used them, as well as coats made out of lawn furniture fabric, but I like nylon ones best—they're light, so the wool doesn't felt underneath; they're puffy, so the wool doesn't press; and they shed water, so the wool doesn't rot.

"The ones I'm using are fabulous! I get them from mohalefarms.com. I've never seen my coated sheep pant any more than non-coatedsheep, and if I stick my hand under their coats, the sheep aren't sweaty. I can't recommend them highly enough. The only drawback is that they do tear easily. One good snag against a fence post splinter and RIIIIIPPPPP, a nice big tear. That's why I get sheep coat patches from Melissa Gray (grayjoe@charter.net); then they last me for years."

—Laurie Andreacci

High-Value Wool

"I did a lot of looking via the Internet and discovered that there is an up-and-coming market for fiber animals to meet the needs of the resurgence of spinning and weaving. There is a demand for both the animals and the fiber.

"Based on the wool sale results at the 2002 Black Sheep Festival in Eugene, Oregon, I narrowed my search down to breeds who have a high-value wool. My next criterion was breeds with lambs selling for a minimum of two hundred dollars a head for breeding purposes. I wanted breeds that would average one hundred dollars a head on the lambs between the ones who went for breeding and the ones sold for meat. My third criterion, since there was going to be a percentage of meat animals raised, was that the breed had to have some size to it. My last criterion was a breed that shorn a minimum of 10 pounds of wool a year."

—Lynn Wilkins

These wool samples are (from left to right) Rambouillet, Border Leicester, Keyrrey-Shee, and Scottish Blackface sheep.

mined by its *grade* (the fineness of the fiber), *class* (its staple length, meaning the length of the fiber), and *quality* (its freedom from contaminants and the character of the fiber).

Fiber diameter is measured in microns (a micron is 1/1,000 of a millimeter, and about 1/25,000 of an inch; this is called the Micron System) or *numerical count* (also called the English or Bradford Spinning Count System). Numerical count refers to the number of 560-yard hanks of wool that could be spun from 1 pound of clean wool. For example, Cheviots grade 27 to 33 microns and 46s to 56s (46 to 56 hanks of yarn).

The American Blood Count System is a largely obsolete third way to measure fiber thickness. In it, wool is compared to Merino fleece, which is considered fine wool (2.5-inch staple with lots of teensy crimps per inch). The seven accepted grades are: 1/2 blood, 3/8 blood, 1/4 blood, low 1/4 blood, common, and braid.

Staple length is the length of a lock (staple) of greasy (unwashed) wool measured in inches in the United States and in millimeters abroad. *Staple strength* is the force needed to break a staple measured in newtons per kilotex. *Crimp* is measured in crimps per inch in the United States and per centimeter elsewhere. Fine wool measures more crimps per inch than do coarser wools. Yarn spun from closely crimped fiber is more elastic than yarn spun from uncrimped wool.

Luster (lustre in Great Britain) equals sheen. Coarser wools from breeds such as the Leicester Longwool, Teeswater, and Wensleydale are more lustrous than wool from fine-wool and down breeds. Sheep coats keep junk such as dirt, burrs, grain, hay, and bedding out of the best parts of a fleece. Although coats don't completely cover a sheep, the exposed wool is on parts *skirted* (cut) off of handspinners' fleeces anyway. *Yield* is how much fleece is left after skirting. Wise producers skirt gen-

It's time for shearing! The quality of fleece—and the price it fetches at market—depends on shepherds' preventive care and maintenance. (See Producing Better Fleeces sidebar.)

erously, removing all low-grade and contaminated wool. Skirted material isn't waste; it can be packaged separately and sold at a lower price to felt makers. *Grease wool* is unwashed fleece, just as it comes off the sheep. Some breeds are greasier than others; Merinos and Rambouillet produce heavy-grease wool. After *scouring* (washing), a fleece weighs considerably less than when it was "in the grease" (unwashed).

Handspinning

Handspinning as a craft has taken the country by storm. If you're interested in producing handspinners' fleeces, there's a ready market for a quality product. The best way to get a feel for handspinning is to try it yourself. Start with simple drop-spindle spinning and if you like it, take some classes and buy a spinning wheel. Due to the burgeoning interest in handspinning, classes are increasingly easy to find. You may want to consider buying a copy of Paula Simmons' *Turning Wool into a Cottage Industry*—it's an absolute must-read.

Drop-Spindle Spinning

As many as 10,000 years ago, even before they domesticated sheep, New Stone Age humans gathered bits of shed wool and spun them into yarn. Archaeologists working Neolithic and Bronze Age digs throughout the world uncover stone whorls from ancient drop spindles.

All wool was processed with drop spindles until the spinning wheel was invented during the Middle Ages. But the drop spindle never fell out of vogue. It's inexpensive, portable, and easy to use. Drop-spindle spinning is still the mode of choice in many third-world countries, and it's enjoying a revival right here at home, too.

CHAPTER EIGHT

Mutton or Milk?

Humankind has dined on lamb and mutton since time began. Artifacts from China's New Stone Age (4000–3000 BC) indicate mutton replaced venison as China's primary meat source by that early date, and archaeologists recovered the bones of at least 500,000 sheep from a 2,600-year-old ritual feasting area on England's Salisbury Plain.

Early humans milked sheep, too. A Mesopotamian clay tablet (circa 2000 BC) documents a farmer's production of sheep's milk, butter, and cheese, proving sheep dairying was established by then.

When most folks think of milk, they rarely think of sheep—but they should. Over 100 million ewes are milked worldwide, particularly in France, Italy, Greece, Turkey, Israel, and Eastern Europe. France alone has more than one million ewes in dairy production. Lamb, mutton, and sheep's milk products remain dietary staples throughout today's world. Shrewd hobby farmers can turn a tidy profit producing them for America's burgeoning lamb and sheep cheese markets.

LAMB AND MUTTON PRODUCTION

Many new shepherds expect to turn a profit selling lamb. Being able to do so means finding the right market and the right breed. Below are some tips to help point you in the right direction.

MARKETS

Compared with citizens of other nations, John Q. Average American eats very little lamb—less than one pound per year, in fact—and almost no mutton (mutton is the stronger-flavored flesh of more mature sheep). One segment of the American population prefers lamb, however, and would like to eat more of it: our rapidly expanding ethnic community. Another lucrative way to earn hobby-farm income with sheep is to niche market grass-fed, natural, or organic lamb.

Ethnic Community: According to the USDA Marketing Service, 75 percent of the lamb meat sold in the United States is imported from Australia and New Zealand. U.S. International Trade Commission figures show a dramatic upswing in lamb imports: 108 million pounds in 2001, up from sixty million pounds just three years earlier. This increase can be attributed to a recent influx of immigrants from particular areas of the globe.

Cabbage leaves are stuffed with ground lamb. Variations on this recipe can be found in Greek, Hungarian, Polish, and Russian cuisines.

The USDA's *Trends in U.S. Sheep Industry* says, "Lamb as a staple seems more typical of Middle Eastern, African, Latin American, and Caribbean consumers. Consumption has remained constant within these groups. The typical lamb consumer is an older, relatively well-established ethnic minority who lives in a metropolitan area such as New York, Boston, or Los Angeles on the West Coast, and prefers to eat only certain lamb cuts." Producing lamb for ethnic consumers is a best-bet proposition for small-scale sheep producers. An especially lucrative market exists prior to holidays. These include Western (Roman) and Eastern (Greek Orthodox) Easter, Christmas, Passover, and the Muslim observances of Ramadan, Eid-al-Fitr, and Eid-al-Adha. However, producers must research each community and market the sort of lamb it requires. The heavy, fatty American market lamb often isn't right.

Lamb and Mutton

Country	Consumed per capita/per annum*
New Zealand	39.6 lb.
Kuwait	38.9 lb.
Australia	35 lb.
Greece	31 lb.
Uruguay1	8.9 lb.
Ireland	14.8 lb.
United Kingdom	14.1 lb.
Saudi Arabia1	3.4 lb.
Spain	13.2 lb.
Bulgaria	9.5 lb.
United States	< 1 lb.

* Susan Schoenian, Sheep and Goat Specialist for the University of Maryland Cooperative Extension, at her www.Sheep101.com Web site.

Delicious lamb chops are cooked to perfection, garnished, and ready to eat. Specialty lamb meat is subject to strict government-mandated regulations, which explains its hefty price tag.

Specialty Lamb: Health-conscious buyers are willing to pay a considerable premium for specialty meat. Of the three types, organic lamb generally fetches the highest prices—but it's notoriously difficult to produce. To call your lamb "organic," you must be enrolled in the United States Department of Agriculture's Certified Organic Program, and qualification is tough. Every nibble of feed and grass your sheep eat must be certified organically grown. Synthetic dewormers are verboten; antibiotics and synthetic medication (other than vaccinations) are taboo. To qualify for organic certification, you must keep detailed flock records and conform to specific handling practices. If a sheep becomes ill, you are required to provide adequate treatment, but if synthet-

ics are used, upon recovery the sheep must be removed from your certified flock. During processing, certified organic lamb can't come in contact with non-organically produced meat, nor may synthetic coloring, preservatives, flavoring, texturizers, or emulsifiers be added to certified organic meat.

Did Ewe Know?

Besides holiday consumption, the birth of Muslim babies often warrants a lamb feast, as do weddings, birthdays, and wakes. Of an estimated five to six million North American Muslims, roughly 2.5 million are converts from other faiths. This trend, coupled with the reality of ongoing immigration, contributes greatly to rapid expansion of the Muslim population on this continent.

Customers crowd an outdoor deli counter that sells fresh sandwiches, specialty meats and cheeses. You can't compete with vendors who offer lamb year-round, but you can turn a tidy profit if you market your meat prior to certain religious holidays.

A less exacting product is "natural" lamb. Natural lamb is generally raised without the use of antibiotics, growth hormones, or stimulants. It is usually grass fed, although some producers feed animals byproduct-free grain for a few weeks or months prior to slaughter.

Grass-fed lamb is, well, grass fed. That is, lambs remain on pasture (and sometimes on stored forage such as hay or silage) from birth to market. Like the others, they're raised without growth hormones. Grass-fed lamb is considerably leaner than commercial supermarket lamb, and it's consistently tested higher in heart-healthy CLA (conjugated linoleic acid).

Selling Methods and Restrictions: Lambs can be sold at auction prior to religious observances, marketed through an intermediary such as a broker, or—as in many cases—direct marketed straight from farm to buyer. If you market lamb from your farm, you pretty much have to sell it on the hoof; federal law prohibits direct-from-the-farm sales of home-processed meat. In most cases, buyers take their live lamb with them; they'll slaughter it themselves or take it to a butcher trained in their religion's beliefs. Some producers allow buyers to process lambs on the farm. Before you do it, check local and state laws; it may not be legal where you live, and to avoid federal recriminations, you should in no way help buyers process their lamb.

To help buyers find you:
• Place ads in metro area newspapers as well as minority publications serving

Religious Observances

LAMB FOR RELIGIOUS OBSERVANCES

Observance	Lamb Requirements
Christmas	Fat, freshly weaned, milk fed, 30-45 lb, not more than 3 months old
Eid-al-Adha	Fairly lean, unblemished lamb (in some circles wethers are unaccceptable, as are docked tails) under one year of age and around 60-80 lb
Western Easter (Roman)	Fat, freshly weaned, milk fed, 30-45 lb, not more than 3 months old
Eastern Easter (Greek Orthodox)	Fat, freshly weaned, milk fed, 40-55 lb, not more than 3 months old
Passover	Freshly weaned, milk fed, 30-45 lb, not more than 3 months old
Start of Ramadan	Under one year of age, fairly lean, around 60-80 lb
Id-al-Fitr	Under one year of age, fairly lean, around 60-80 lb

your target group within that locale.

- Post flyers and business cards in ethnic neighborhoods and at farmer's markets.
- Contact TV and radio stations catering to your target group. They're always seeking newsworthy stories, and they might use yours.
- A month or so before a religious holiday, phone mainstream newspapers and TV stations. Your human interest story may be precisely what they're seeking.

As with lambs for the ethnic markets, unless you're licensed to sell processed meat, you must sell live lambs for the specialty market. Most producers price their lambs delivered to a USDA-approved slaughtering facility. Customers pick up the packaged meat themselves.

These Barbados Blackbelly crossbreds—a primitive flock with beautiful camouflage—are a good choice for hobby farmers who want to produce better-tasting meat.

THE RIGHT BREED

If you're eager to enter the ethnic or specialty lamb meat market but haven't settled on a breed, think hair sheep. In North America, this means Katahdins, St. Croix, Barbados Blackbellies, Dorpers, and Wiltshire Horns. Some have hair coats, and others shed their wool, but none of these sheep have to be shorn.

To various degrees, all hair-sheep breeds are resistant to internal parasites and foot rot, tolerant of heat, and uncommonly productive. They are peerless foragers and browse more than wooled sheep. All breed out of season; with proper care, they can lamb three times in two years. They're attentive mothers, give oodles of milk, and generally produce triplets or twins. According to university studies in Georgia, California, and Mississippi, hair-sheep flesh is milder, thus tastier, than that of wooled sheep lamb and mutton. Some ethnic minorities greatly prefer hair-sheep lambs and will pay premium prices for them even when auctioned.

MILK: THE NUTRITIOUS STUFF

At 5.6 percent, sheep's milk is higher in protein than cow or goat milk (3.4 percent and 2.9 percent, respectively), and it packs up to twice as much calcium, phosphorus, and zinc. It also boasts a

Though traditionally made with cow's milk, varieties of sheep's milk-based brie (above) are becoming increasingly popular. Le Berger de Rocastin, for example, has a creamy interior and edible rind and ripens in just a few weeks.

hefty 6-percent fat content and contains considerably more solids than other dairy milks, making it an ideal medium for crafting luscious, full-bodied cheeses.

Because sheep cheese breaks down into smaller molecules than other types, it's more easily digested. Lactose-intolerant diners often enjoy sheep's milk cheese. Its hearty flavor quickly satisfies the palate; less is eaten, so less fat is consumed per sitting.

For these reasons, and because it just tastes so good, the United States imports more than seventy-five million pounds of sheep's milk cheese per year. In 1995, Canada imported more than two million kilos of the stuff. You've probably eaten it; sheep's milk cheeses include Roquefort, Ossau-Iraty-Brebis, and Le Berger de Rocastin (sheep's milk brie) from France; Brin D'Amour from Corsica; Roncal, Zamarano, Iberico, and Manchego from Spain; Greece's feta, Kasseri, and Manouri; Pepato and Ricotta Salata from Sicily; and Italy's Pecorino, including word-famous Romano.

Now American artisan and small-scale cheese makers are turning to sheep cheese in droves. To supply their demand for raw product, more sheep dairies are flocking to the fold each year. The need will only increase.

THE ADVANTAGES OF PRODUCING SHEEP'S MILK

There are many advantages to producing sheep's milk. It costs less to equip a brand-new sheep dairy than a comparable cattle operation, and existing cattle facilities can be converted to sheep dairying at relatively low cost. Quality dairy ewes cost less to buy and maintain than cows of the same caliber. They are also easier to handle and less likely to damage costly equipment.

Where a market exists, sheep's milk generally sells for about four times the price of cow milk. Sheep's milk freezes

(Ewe) Got Milk?

SHEEP'S MILK VERSUS GOAT AND COW MILK

	Sheep	Goat	Cow
Whole Milk %			
Total Solids	18.3	11.2	12.1
Fat	6.7	3.9	3.5
Protein	5.6	2.9	3.4
Lactose	4.8	4.1	4.5
Calories/100g	102.0	77.0	73.0
Vitamins mg/l			
Riboflavin B2	4.3	1.4	2.2
Thiamine	1.2	0.5	0.5
Niacin B1	5.4	2.5	1.0
Pantothenic acid	5.3	3.6	3.4
B6	0.7	0.6	0.6
Folic acid ug/l	0.5	0.06	0.5
B12	0.09	0.007	0.03
Biotin	5.0	4.0	1.7
Minerals mg/100g			
Calcium (Ca)	162–259	102–203	100–120
Phosphorous (P)	82–183	86–118	90–90
Sodium (Na)	41–132	35–65	56–60
Magnesium (Mg)	14–19	12–20	10–12
Zinc (Zn)	0.5–1.5	0.18–0.5	0.2–0.4
Iron (Fe)	0.02–0.01	0.01–0.1	0.02–0.06

well for up to six months; it can be stockpiled until the supply warrants shipment.

Ewe milk registers 18.3 percent solid content, as opposed to cow milk's 12.1 percent. It takes ten pounds of cow milk—and only six of sheep's milk—to craft a one-pound block of cheese.

But there are disadvantages, too. Production of sheep's milk cheese is still a developing market in North America, and processors are few and far between. According to Penn State's Milking Sheep Production, most processors require a steady supply of milk from at least 750 ewes. Producers may need to form sheep dairy co-ops to meet demands or produce value-added products, such as cheese or yogurt, and market it themselves. Due to stringent North American livestock importation regulations, purebred European dairy sheep are relatively expensive—and scarce. Where a market exists—or if you're willing to create one—sheep dairying can be a viable small-farm moneymaker. But it's not for everyone.

Roquefort is one of several gourmet cheeses made from sheep's milk. These hearty flavored dairy delights are more easily tolerated by the lactose-intolerant.

Talk to people in the business—lots of them. Subscribe to the DairySheep listserv. Join sheep dairying organizations and attend their functions. Read everything you can lay your hands on, including online resources (see Appendix). Do the homework carefully before you commit.

HOUSEHOLD DAIRY SHEEP

Sheep make fine kitchen dairy animals, too. Depending on their breed and individual productivity, two to five ewes can keep a typical family in luscious milk, cheese, and yogurt year-round. But before you buy any dairy animal, be absolutely certain you want one. Milking ties you down. Someone has to milk the ewes twice a day, every day, at the same time—no exceptions—for their entire three to seven month lactation. The milk has to be strained, cooled, and possibly bagged and frozen, adding another block of time to the daily equation. If you get sick or hurt, or are called away on urgent business, you'll have to find someone to field for you. If that doesn't sound like your cup of tea, buy your milk at the supermarket.

To produce milk, your ewes must lamb every year. Are you willing to keep a ram or find and pay for stud service for your girls? Hauling them to another

The Rambouillet ewe at right could make a great home dairy sheep. In the right market, the profits from selling sheep's milk can be good, but the cost in time and labor is considerable for a small-scale operation.

farm could expose them to disease, as could bringing a ram to them. Will you keep the lambs? Sell them? Eat them? Will you milk with a machine, or the old way, by hand? Expect to pay in the neighborhood of fifteen hundred dollars for a complete belly-pail milking system for one ewe, or sixteen hundred to eighteen hundred dollars for a two-ewe setup.

Hand milking a gentle ewe or two can be the most relaxing, meditative time of your day, however, and it's easier than you think. If you've never milked before, start with trained dairy ewes or other docile sheep that are accustomed to being handled. Battling flighty, rambunctious, "don't touch me!" ewes twice a day will burn you

out fast. If production isn't an issue, any easygoing ewe you already own can be milked, if she's trained or easily trainable. She may not give a lot, but compared with supermarket cow milk, it's bound to be mighty good!

If you'd like to hand milk a ewe or two or three, start your miling string with newborn lambs. Choose ewe lambs from productive, "milky" mamas, then bottle-raise them so they grow up tame. Halter your future milkers; teach them to lead and stand patiently tied. Pet them a lot, and massage their udders so they accept your touch. When they're old enough to lamb, they'll be ready-made milkers. It's the easiest way to train milk sheep.

These St. Croix hair sheep possess a number of positive attributes: they make excellent dairy sheep, don't require shearing, show high resistance to certain internal parasites, and can tolerate hot weather.

DAIRY-SHEEP BREEDS

Whether you milk two sheep or two hundred, choosing ewes from proven dairy stock is time and cost effective. Buy from a reliable producer. Many quality dairy ewes are crossbreds, so registration papers won't matter unless you breed registered lambs. Detailed health and production records, however, are essential.

East Friesian, Lacaune, and British Milk Sheep ewes are the pros of the dairy-sheep world; they give four hundred to one thousand pounds (or more) of milk per 120- to 240-day lactation. Heavy milking non-dairy breeds produce one hundred to two hundred pounds of milk in 90 to 150 days. Crossbreds strike a happy medium at 250–650 pounds per lactation, and they cost far less than their purebred sisters do.

*Heavy Milking Non Dairy-Specific Ewes
(from Penn State's bulletin *Milking Sheep Production*)

- Coopworth
- Rambouillet
- Dorset
- Romanov
- Finnsheep
- Shropshire
- Katahdin
- St. Croix
- Polypay
- Targhee

* *averaging 120–150 lb milk per lactation*

Another promising dairy sheep is the purebred Icelandic. In addition to her milk-producing capabilities, she's a beautiful animal. Icelandic fleece is coveted by and readily marketed to handspinners, and registered Icelandic breeding-stock lambs are in high demand.

Acknowledgments

Thanks again to the good folks who contributed "Advice from the Farm" tips and words of wisdom.

Laurie Andreacci lives in Chesterfield, New Jersey. She raises Shetlands and Tunis and is bringing Gotland Sheep to the United States. Contact Laurie at Laurie's Lambs, (609) 324-0487 <http://www.laurieslambs.com>

Cathy Bridges is a third-generation sheep farmer who lives near Silverton, Oregon. Her commercial flock consists of crossbred ewes bred for heartiness, milk, and mothering skills. Cathy also keeps a natural-colored Romney flock for handspinners. Contact Cathy at CathyAndEwe@ CathyBridges.com.

Jerry and Lyn Brown raise Registered California Red Sheep in the La Plata Valley of the Four Corners area of New Mexico. Lyn devotes many hours each year to organizing and promoting the Wool Festival of the Southwest <http://www.woolfestivalsw.meridian1.net>. Contact Lyn at <http://www.nmredsheep. meridian1.net> or nmredsheep@yahoo.com.

Barbara Burrows raises colored Rambouillets and is one of only a few American breeders of Teeswater and Wensleydale sheep. (Tremont and Edmund are wether lambs from Barbara's flock; she has fleece lambs like our boys for sale every year.) Contact Barbara at Ewes in Color <http://myweb.wyoming.com/~bburrows> or ewesincolor@wyoming.com.

Melissa Gray raises Corriedale sheep and sells sheep coat patches and Baa Bars sheep treats through her home enterprise, ABC Woolcrafts <http://abcwoolcrafts.mralter.com>. Contact Melissa at grayjoe@ charter.net.

Bernadette McBride lives in Andover, Connecticut, with her four young children and husband, John. She raises Traditional Siamese and Balinese cats and recently acquired a pair of miniature Cheviot sheep named Fiona and Angus. She can be reached at bernadette.mcbride@comcast.net.

Lisa MacIver is an experienced shepherd and goat fancier. She resides in Ashland, Oregon.

Kim McCary is a former shepherd living in Waverly, Ohio. Kim is presently involved with dairy goats.

John Weaver is the author's husband and co-photographer for this book. Enough said!

Connie and Amy Wheeler are a mother-daughter team in Molalla, Oregon. They raise St. Croix, Katahdin, DorpKats, and Wool-Hybrid Sheep. Contact Connie and Amy at Hollow Hills Ranch <http://www.molalla.net/~amylynn/farm/farm.html>, (503) 829-8148, or hollow_hills_ranch@molalla.net.

Lynn Wilkins lives near Condon, Oregon, and manages the family cattle ranch with her husband. She raises registered rare breed fiber sheep. She can be reached at (541) 384-3699 or at lwilkins6@yahoo.com.

Appendix:
A Glance at
Sheep Afflictions

ABORTION

- Enzootic abortion of ewes (EAE) is a chlamydial disease transmitted from aborting sheep and fetal tissues to other ewes. Infected ewes abort during the last month of pregnancy or give birth to stillborn or weak lambs that soon die. An effective vaccine is available.

- Vibrosis is caused by the bacterium *Campylobacter fetus*, subspecies intestinalis. When one or two ewes affected by Vibrosis abort, they can trigger an "abortion storm." Vibrosis vaccine is available, often in combination with EAE vaccine.

- Toxoplasmosis, which is caused by the coccidium *Toxoplasma gondii*, is spread when a host cat contaminates feed and water with his droppings. There is no vaccination or treatment for toxoplasmosis.

- When a ewe aborts her lamb, the fetus and tissues should be submitted to a laboratory for diagnosis; you can't treat the rest of the flock unless you know what you're dealing with. Your vet can tell you where to send the specimens. The material must be fresh, so store it in sturdy plastic bags, pack the bags in a Styrofoam box and surround them with chill packs, then rush the package to the lab as fast as you can.

BLACKLEG AND MALIGNANT EDEMA

- These deadly diseases, caused by the bacteria *Clostridium chauvoei* and *Clostridium septicum*, respectively, are indistinguishable in sheep except by laboratory diagnosis.

- They occur in sheep of all ages and are caused when soil-borne bacterium contaminate wounds and abrasions.

- A combination vaccine prevents most occurrences.

BLOAT

- Bloat is a build-up of frothy gas in the rumen.

- Bloat is usually triggered when a sheep tanks up on an unaccustomed abundance of grain, rich grass, or legume hay.

- Bloated sheep can quickly die of the condition, so if you suspect that your sheep has it, call your vet posthaste.

BLUETONGUE

- Bluetongue is caused by reoviruses transmitted by *Culicoides variipennis* gnats. It infects sheep, goats, cattle, and wild ruminants such as deer, elk, and moose.

- The classical form of bluetongue occurs in sheep, causing fever and swelling of the lips, tongue, and gums. Badly chewed tongues turn purple due to lack of oxygen, hence the name bluetongue. Afflicted sheep have difficulty breathing and swallowing.

- Death usually occurs in about seven days, although some sheep survive.

CASEOUS LYMPHADENITIS (CLA)

- Caseous lymphadenitis (CLA) is a chronic, contagious disease of sheep and goats caused by the bacterium *Corynebacterium pseudotuberculosis*. The bacterium breaches a sheep's body through mucous membranes or via cuts and abrasions. The animal's immune system valiantly tries to localize the infection by surrounding it in one or more cysts. If the ploy is unsuccessful, she will die.

- CLA presents as lumps near the jaw, in front of the shoulder, and where her udder attaches to her body. Some sheep develop internal cysts too.

CONTAGIOUS ECTHYMA

- Commonly known as sore mouth—also known as scabby mouth, or orf—contagious ecthyma (CE) is a contagious poxlike virus that causes the formation of blisters and pustules on the lips and inside the mouths of young lambs, as well as on the teats of the infected lambs' mothers. The blisters pop, causing scabbing and pain so intense that a lamb will occasionally starve rather than eat. Most lambs recover in one to three weeks without treatment.

- Because sore mouth is easily transmissible to humans, you should wear rubber gloves when handling stricken lambs. Keep children away from all infected sheep!

- An effective live vaccine is available, but you mustn't use it unless you already have sore mouth on your property; when live vaccination scabs fall off your inoculated sheep, you will certainly have it.

ENTEROTOXEMIA

- There are two types of enterotoxemia in sheep: *Clostridium perfringens* C and D.

- The first is a disease of young lambs caused by *Clostridium perfringens* type C, an anaerobic bacterium found in manure and soil. It enters via newborn lambs' mouths when they suckle dirty wool or manure tags while seeking their mothers' udders. Bacteria produce a toxin that causes rapid death. Treatment is unsuccessful, but lambs from ewes vaccinated for enterotoxemia during late pregnancy develop immunity via their mothers' colostrum.

- The second form is caused by *Clostridium perfringens* type D, also present in the soil and manure. It attacks rapidly growing, slightly older lambs that ingest the bacterium while investigating their environment. It too causes rapid death and with it tremors, convulsions, and a host of strange neurological behaviors. Vaccine is available alone or in combination with type C or type C and tetanus vaccine.

JOHNE'S DISEASE

- Johne's (YO-nees) is a deadly and contagious, slow-developing, antibiotic-resistant disease affecting the intestinal tracts of domestic and wild ruminants, including sheep.

- The bacterium that causes Johne's, *Mycobacterium avium subsp.* paratuberculosis, is closely related to the one causing tuberculosis in humans. Infected sheep are dull, depressed, and thin. Johne's disease, also known as Paratuberculosis, is incurable.

MASTITIS

- Mastitis is an inflammation of the udder caused by bacteria or yeast infections.

- Usually only one side is affected. One type of acute mastitis can lead to gangrene.

- Anyone who keeps brood stock or milking ewes should learn to recognize mastitis symptoms.

OVINE PROGRESSIVE PNEUMONIA

- OPP is one of the serious, progressive diseases shepherds fear. According to a recent study, 26% of America's sheep are infected.

- This North American retrovirus is closely related to *maedi-visna*, a similar disease found in most other parts of the world. Only Iceland, Australia, and New Zealand are *maedi-visna* and OPP free.

- Transmission is via infected colostrum and milk, but because of the disease's slow progression, infected sheep do not develop symptoms until much later. Symptoms include weight loss, sluggishness, lameness, a fibrous udder condition known as *hardbag*, and all the usual signs of pneumonia.

- Most OPP-infected sheep develop secondary bacterial pneumonia, from which they die.

- There is no vaccine or treatment for OPP.

PNEUMONIA
- Pneumonia is caused when one of a wide variety of opportunistic bacteria and viruses mix with stressed sheep.

- Typical symptoms include depression, fever, coughing, and labored breathing. Because so many bacteria and viruses may be involved, accurate identification of the infectious agent is an essential part of successful treatment.

SCRAPIE
- Scrapie (scray-pee) is a transmissible spongiform encephalopathy (TSE) similar to bovine spongiform encephalopathy (BSE, or Mad Cow Disease) and Chronic Wasting Disease (CWD, which affects deer and elk). No human has ever contracted scrapie (or either of the human equivalents, Kuru and Creutzfeldt-Jakob disease) from sheep.

- As a sheep owner of the new millennium, you must become scrapie savvy—it's the law. All sheep residing within the United States must be identified through one of two government scrapie programs—one of which is voluntary; the other, mandatory. At press time, the provisions of both programs are rapidly changing, so contact your state Animal and Plant Health Inspection Service/ APHIS representative for up-to-date information.

- Scrapie appears to be caused by an infectious agent, but genetics also play a part. The disease was recognized in Britain and western Europe at least 200 years ago, and it came to the United States in 1947 with British sheep. Scrapie is a global scourge: only Australia and New Zealand are scrapie free.

- Scrapie is a slow, progressive disease that systematically destroys the central nervous system. Goats exposed to infected sheep sometimes get scrapie, too. Symptoms typically appear two to five years after infection and include weight loss, tremors, bizarre locomotion (such as bunny-hopping or high-stepping like a Hackney horse), swaying, stumbling, wool pulling, lip smacking, and intense itchiness. Between one and six months after symptoms appear, infected sheep die.

URINARY CALCULI
- Urinary calculi are tiny stones or crystals that form in the urinary tracts

of sheep and goats. Ewes get stones, but they pass through the larger female urethra (the tube that empties urine from the bladder) without difficulty. A ram or wether with a blocked urethra is in trouble, however; his bladder is apt to rupture, and he'll probably die.

- When ram lambs are castrated, penis growth stops, so wethers castrated early are especially troubled by calculi; their much tinier penises and urethras are easily blocked. A workable solution: don't castrate ram lambs younger than four to six weeks old.

- A calcium-phosphorus ratio of 2 to 1 in the diet helps prevent calculi formation, as do minute amounts of ammonium chloride added to feed. Male sheep should drink lots of water. Make it more appealing by keeping water sources readily available, full, and sparkling clean.

WHITE MUSCLE DISEASE

- White muscle disease, also known as stiff lamb disease, is caused by selenium deficiency.

- Ewes grazing on selenium-poor land or eating hay raised in such depleted conditions require selenium/vitamin D supplementation during the last two months of pregnancy. Otherwise, their affected lambs will have problems rising and walking. Some will even become paralyzed. Prevention is the key to eliminating white muscle disease.

Glossary

Abomasum—the fourth compartment of a ruminant's stomach

A. I.—artificial insemination

Band—(noun) a strong rubber band used to dock and castrate lambs; (verb) the act of using an Elastrator to apply one of these bands

Black-face sheep—breeds raised mostly for meat

Body condition score—a rating of 1 to 5 (very thin to obese) used to estimate the condition of sheep

Breech birth or breech delivery—one in which a lamb's hind feet come first

Broken mouth—an old sheep with missing or broken teeth

Browse—(noun) edible woody plants such as twigs or saplings and wild berry canes; (verb) the act of eating browse

Bummer—a lamb that sneaks milk from a ewe other than its dam; also used to denote a bottle-fed lamb

Burdizzo—a tool used to castrate lambs by severing the cord without breaking the skin of the scrotum

Card—(noun) a wire-toothed tool used to untangle wool; (verb) the act of untangling wool with a card

Clean legged—clean legged sheep have little or no wool growing on their legs

Clip—one season's wool harvest

Closed face—a sheep with heavy wool surrounding its eyes and covering its cheeks

Club lamb—a lamb raised as a 4-H or an FFA project

Colostrum—a ewe's first milk; it contains antibodies that protect her lambs through their first few months of life until they develop disease resistance of their own

Coronary band—the soft area where hoof meets leg

Count System—(also called the Bradford or English Spinning Count System)—a system for grading wool based on how many

560-yard hanks of single-ply yarn can be spun from one pound of wool; the higher the count, the finer the wool

Creep—a feeder designed so lambs can enter and eat but larger sheep cannot

Crimp—the natural waviness of wool fiber that allows it to stretch and **then spring back**

Crossbred—an animal with parents of two different breeds

Crutching—(verb) the act of shearing wool from between a sheep's back **legs and belly; (noun) a hunk of material removed while crutching**

Cud—a glob of regurgitated food that's re-chewed and swallowed again

Cull—(verb) the act of removing undesirable sheep from a flock; (noun) a sheep removed as part of the culling process

Dam—an animal's female parent

Dental pad—a sheep's firm upper palate

Docking—the act of shortening a lamb's tail

Drench—(noun) liquid medicine given orally; (verb) to administer a drench

Dual-coat—the type of fleece produced by certain primitive breeds such as the Shetland, Icelandic, and Navajo-Churro, consisting of a longer outer coat and a distinctively different, shorter, fluffier undercoat

Dual-purpose breeds—breeds developed for both wool and meat production

Elastrator—a tool used to apply heavy rubber bands to lambs' tails or scrotums for docking or castration

Emasculator—a tool used for docking and castrating lambs; it has a crushing effect, which helps reduce bleeding

Estrus—heat; the period during which a ewe is receptive and can conceive

Ewe—a mature female sheep

Ewe lamb—a female lamb

Facing—the act of trimming wool away from the eyes of a closed-faced sheep

Fine wool—soft wool with small-diameter fibers

Fleece—raw wool, usually in one piece, as shorn from a single sheep

Flock—a group of sheep

Flushing—increasing a ewe's nutritional level prior to breeding season

Fly-strike—a condition caused when blowflies lay eggs in wounds or wet, filthy fleece; maggots develop and consume the host's flesh

Foot bath—a chemical mixture sheep walk through or stand in, designed to prevent or treat hoof disease

Forage—fibrous animal feeds such as grass and hay

Free choice—food available 24/7

Gestation—the period of pregnancy beginning at conception and ending at birth

Gimmer—a ewe lamb between weaning age and first shearing

Grade—an unregistered sheep

Graft—the act of persuading a ewe to adopt another ewe's lamb or lambs

Guard dog (llama, donkey)—an animal that bonds with and stays with sheep to guard them from predators such as coyotes, dogs, wolves, bears, cougars, and eagles

Gummer—an old sheep with no teeth

Halter—headgear used to lead or tie an animal

Handle—how a fleece feels to a spinner

Heat—estrus; the period when a ewe is receptive to a ram and can conceive

Hogget—a young sheep between weaning age and first or second shearing, depending on locale; also called a hog, hogg, or hoggie

Hybrid vigor—the extra strength, hardiness, and productivity exhibited by animals whose parents are of two different breeds

Jug—a cozy pen used by a single ewe and her newborn lambs, where she can bond with and watch over them in peace

Ketones—compounds found in the blood of pregnant ewes suffering from pregnancy disease (ketosis)

Lactation—the period during which a ewe produces milk

Lamb—(noun) a baby sheep, a sheep less than one year old; (verb) the act of giving birth to lambs

Lamb fleece—a young sheep's first fleece, usually the softest and finest it will produce

Lanolin—naturally occurring grease in wool

Linebreeding—the breeding of closely related sheep; used to fix type and intensify the characteristics of shared ancestors

Liver flukes—tiny leaf-shaped parasites that dwell in bile ducts and liver tissue

Lungworms—parasites that infest the respiratory tract and lung tissue

Luster—sheen caused by lanolin in fleece

Marking harness—a harness apparatus containing a colored crayon or colored chalk, worn by rams during breeding season

Mastitis—inflammation of the udder

Mating capacity—the number of ewes a ram can impregnate in a season

Meconium—the first manure passed by a newborn lamb

Micron System—a system by which wool fiber is measured in microns (1/1000 of a millimeter or about 1/25,000 of an inch)

Mutton—the meat of adult sheep

Open-faced—a sheep with little or no wool on its face

Ovine—referring to sheep

Ovulation—the period when an egg is released from an ovary and a ewe can conceive

Oxytocin—the hormone that controls milk letdown; oxytocin shots are sometimes given to help ewes expel afterbirth tissue

Papered—registered

Pasture lambing—the act of allowing ewes to give birth at pasture instead of in a barn

Pelt—a wooled sheep hide

Polled—naturally hornless

Purebred—an animal whose ancestors for several generations were all of the same breed

Ram—an uncastrated adult male sheep

Ram lamb—an uncastrated male lamb

Raw wool—unwashed wool

Registered—a purebred animal whose pedigree and particulars are recorded in a registry's official flock book

Roving—washed and carded wool, ready for spinning

Rumen—the large, first stomach compartment of a ruminant, where feed is broken down into usable elements

Ruminant—an animal with a four-compartment stomach and who chews cud

Ruminate—the act of chewing cud

S/T/Tr—shorthand for single, twin, and triplet births

Scours—diarrhea

Scrapie—a serious, transmissible spongiform encephalopathy malady much like mad cow disease

Scur—a horn button on an adult sheep or a small, misshapen horn

Second cuts—short pieces of wool on or in a shorn fleece created by reshearing an already-shorn spot

Septicemia—an infection of the bloodstream that affects the entire body

Shearing—the act of removing wool from sheep

Sheep keds—bloodsucking wingless flies; sometimes called sheep ticks

Shepherd—a person of either sex who tends sheep

Sire—an animal's male parent

Shrink—the weight and volume lost through washing wool

Skirt—to remove undesirable parts of a shorn fleece

Stocking rate—the number of animals grazed on an acre of land

Subcutaneous injection (SQ)—an injection inserted directly under the skin

Tag—(noun) a dreadlock of manure-laden wool; (verb) the act of clipping tags from a sheep's fleece

Tapeworms—long, ribbonlike flatworms dwelling in the gastrointestinal tract

Tippy—fleece with twisted or lightly matted fiber tips

Tup—a ram

Urinary calculi—stones formed in the urinary tract

Vaginal prolapse—protrusion of part or all of the vagina in late-gestation ewes

Wether—(noun) a castrated male sheep; (verb) the act of castrating a male sheep

White-face sheep—breeds raised primarily for wool

Wool—sheep fiber

Wool blindness—a condition caused by excessive wool growth around a sheep's eyes

Wool break—weak places in a fleece caused by fever, stress, or other trauma

(Wool) in the grease—unwashed wool

Yearling—a sheep of either sex between one and two years of age

Resources

Main Online Resources

The Internet is a modern shepherd's best friend. Whatever information you seek, it's there—if you know where to look. With that in mind, we're here to guide you to the best sheep-savvy Web sites in the world.

NORTH AMERICAN SHEEP BREEDERS' DIRECTORIES

BREEDERS WORLD SHEEP DIRECTORY
www.breedersworld.com
Breeders World bills itself as "The Complete Online Livestock Directory." Click on the *Sheep* link on their home page to access their Sheep Breeders' Directory, chat rooms and breeders forum, equipment and book suppliers' pages, and a comprehensive list of sheep associations.

DMOZ OPEN DIRECTORY PROJECT
dmoz.org/Business/Agriculture_and_
Forestry/Livestock/Sheep
The Open Directory Project is the largest human-edited directory on the World Wide Web. Close to 1,000 sites are cataloged in their sheep resources directory. Visit to locate sheep associations, breeders, supplies and equipment, sheep shearers, lamb and mutton recipes , and educational sites galore.

YAHOO! DIRECTORY

dir.yahoo.com/Science/Agriculture/
Animal_Science/Sheep
The Yahoo! Sheep Science Directory catalogs scores of university and sheep registry educational sites. A click on this page's *Commercial Livestock link* leads to additional Yahoo! Sheep Directories including: *Breeders, Scrapie Information, Lamb Meat,* and *Wool.*

U.S. SHEEP BREEDERS ONLINE DIRECTORY

www.nebraskasheep.com

The Nebraska Sheep Web site belongs to Drudik Suffolks of Grand Island, Nebraska. A click on the colorful U.S. *Sheep Breeders Online Directory* box leads to the most comprehensive guide to commercial sheep breeders on the World Wide Web. You'll find more than 1,500 breeders listed—including more than 200 suppliers of club (4-H and FFA) lambs—along with shearers, sheep haulers, sheep nutrition resources, and more. While visiting the Drudik home page, click on *Tips and Topics* to access their sheep management article archives and *Sheep Markets* to see how sheep are selling at major livestock markets throughout the United States.

SHEEP ORGANIZATIONS
UNITED STATES
AMERICAN LAMB BOARD/ALB

www.americanlambboard.org

If you prepare lamb for your family's table, this is your site. Nutrition information, how to buy and prepare lamb, and hundreds of great lamb recipes—they're here.

AMERICAN SHEEP INDUSTRY ASSOCIATION/ASI

www.sheepusa.org

ASI is a national organization representing 64,000 sheep producers throughout the United States. Market summaries, legislative news, research and information, and a consumers' section offering lamb recipes, wool-care tips, and a kids' corner are a few of this site's many resources. Don't miss the breed description pages; they're outstanding.

AMERICAN WOOL COUNCIL/AWC

www.americanwool.org

Click on *General Wool Info* to learn the history, care, and characteristics of wool and how it's processed.

CALIFORNIA WOOL GROWERS ASSOCIATION/CWGA

www.woolgrowers.org

The 130-year-old California Wool Growers Association is an active organization of West Coast sheep breeders and producers. Web pages are devoted to current industry issues and market prices, as well as fascinating *Fast Facts* about topics such as *Sheep Ecology in America* and *American Wool* and a collection of tasty lamb recipes. Request a free copy of the CWGA's print newsletter when you visit.

DAIRY SHEEP ASSOCIATION OF NORTH AMERICA/DSANA

www.dsana.org

If you're investigating sheep dairying, don't miss this site! Read about the organization and its annual symposia, take a DSANA sheep dairy virtual tour, and visit the comprehensive links pages where categories include *Sheep Dairy Production, Sheep's milk in the Press, Cheese and Cheesemaker Links, Cheesemaking Links, Farm Links, Animal Health, Human Health, Sheep's milk, Suppliers, Utilities,* and *Wool.* This is a five-star site!

DUTCHESS COUNTY SHEEP AND WOOL GROWERS ASSOCIATION

www.sheepandwool.com

The Dutchess County Sheep and Wool Growers Association site helps sheep fanciers locate breeders of registered and grade sheep, sheep shearers, spinners and weavers, and sheep shearers doing business in New York's Hudson Valley. It also maintains a breeders list, highlights a Sheep Breed of the Year, and sells wonderful New York State Sheep and Wool Festival promotional items. Don't miss it!

GARDEN STATE SHEEP BREEDERS

www.quintillion.com/gssb/index.html

The high point of the Garden State Sheep Breeders Web site is its section on sheep breeds, where virtually every breed available in North America is allotted its own page. You'll also find classifieds, a members directory, and a fun gallery of sheep images at this nicely done site.

KENTUCKY SHEEP AND WOOL PRODUCERS ASSOCIATION

www.kswpa.com

The Kentucky Sheep and Wool Producers Association promotes the sheep industry throughout Kentucky. Pages describing breeds raised in Kentucky and a series of wool-related articles appeal to all sheep fanciers; a breeders directory and classified ads serve area shepherds to a tee.

MAINE SHEEP BREEDERS ASSOCIATION

www.mainesheepbreeders.org

Learn about annual Maine Sheep Breeders Association shepherds' and shearers' schools, peruse an online producers directory, and click on the *Special Topics* link to access a slew of fine educational resources.

MICHIGAN SHEEP BREEDERS ASSOCIATION/MSBA

www.misheep.org

Visit the Michigan Sheep Breeders Association site to scope out current sheep and lamb market prices within the state, catch up on news, and search the comprehensive *MSBA Directory*.

MID-STATES WOOL GROWERS COOPERATIVE ASSOCIATION/ MSWGCA

www.midstateswoolgrowers.com

Formed in 1918, the Mid-States Wool Growers Cooperative serves more than 10,000 shepherd/owners from 23 states. Peruse market reports, marketing and management tips, and up-to-date North American and International wool industry news at this Web site. Click on *Sheep Supplies* to enter the co-op's online store, where you'll find everything you need to raise sheep and then some (request a print copy while you're there—it's free).

MINNESOTA LAMB AND WOOL PRODUCTS ASSOCIATION/ MLWPA

www.mlwp.org

The Minnesota Lamb and Wool Products Association site is a good one. Be sure not to miss the *MLWPA Direct Marketers* ads—a unique listing of Minnesota shepherds specializing in straight-from-the-farm sales of handspinners fleeces, yarn, spinning wheels, tanned lambskins, farm fresh lamb, and more. You can also download a selection of *MLWP Newsletters* from this site.

MISSOURI NATURAL COLORED WOOL GROWERS ASSOCIATION

www.moncwga.com

A *Membership Registry,* ads for animals and equipment, and information about the annual Heart of America Sheep Show and Fiber Fest are highlights of the Missouri Natural Colored Wool Growers Association Web site.

MONTANA WOOL GROWERS ASSOCIATION

www.mtsheep.org

The Montana Wool Growers Association promotes the welfare of sheep and woolgrowers in the state of Montana. The site offers separate sheep shearers and breeders directories, world wool-market reports, and extensive sheep links pages. National Lamb Feeders Organization www.nlfa-sheep.org

The National Lamb Feeders Association sponsors programs and activities to promote and improve the production of lambs and sheep in the United States and throughout the world. Visit the site to download the latest NLFO Newsletter in PDF format.

NATIONAL SHEEP INDUSTRY IMPROVEMENT CENTER/NSIIC

www.nsiic.org

The National Sheep Industry Improvement Center (NSIIC) was established as a revolving fund by the 1996 Farm Bill to aid the nation's ailing sheep and goat industries. Visit the NSIIC site to bone up on the latest sheep industry legislation and learn all about NSIIC grants and loans.

NATURAL COLORED WOOL GROWERS ASSOCIATION/ NCWGA

www.ncwga.org

The Natural Colored Wool Growers Association promotes natural-colored (non-white) sheep and the wool they produce. The organization also maintains a registry for colored sheep of all breeds. Visit the site to read about the history of colored sheep, locate a wool festival in your locale, or download registry materials. Stop by their online store to shop for cool NCWGA promotional items.

NORTH DAKOTA LAMB AND WOOL PRODUCERS ASSOCIATION

www.ndlwpa.com

Member links, North Dakota State University sheep links, and a *Producers Directory* are features of the North Dakota Lamb and Wool Producers Association Web site. Dozens of back issues of the *North Dakota Lamb and Wool Industry Newsletter* are archived for readers' convenience.

TREASURE VALLEY SHEEP PRODUCERS/TVSP

www.tvsp.org

The Idaho-based Treasure Valley Sheep Producers Web site is a treasure trove of sheepy resources. Scope out the *Producers*

Directory and comprehensive marketing news and links pages, and scroll down the TVSP home page to an impressive list of downloadable educational sheep bulletins.

VERMONT SHEEP AND GOAT ASSOCIATION

www.vermontsheep.org
The Vermont Sheep and Goat Association site offers a *Business Directory* and links to a plethora of helpful online sheep materials. Click on *Resources, Animal Health*, and *Library*—and learn about the Vermont Sheep and Wool Festival while you're there.

SHEEP ORGANIZATIONS CANADA
CANADIAN CO-OPERATIVE WOOL GROWERS LIMITED

www.wool.ca
The Canadian Co-operative Wool Growers site features a comprehensive breeders directory, an extensive sheep links feature (click on *Wool Links* in the menu), a completely stocked online sheep supply catalog, and a Real Wool Shop featuring quality items crafted of wool and sheepskin.

CANADIAN SHEEP FEDERATION/CFS

www.cansheep.ca
The Canadian Sheep Federation was established in 1990 to set national policy for the Canadian sheep industry. The CFS Web site is packed with news, market results, loads of articles addressing health and management topics, consumer tips

(including lamb recipes), and links to other Canadian sheep sites and organizations.

ONTARIO DAIRY SHEEP ASSOCIATION/ODSA

www.sheepmilk.com/odsa
If you're interested in sheep dairying you'll find lots to peruse at the Ontario Dairy Sheep Association Web site. Click on *State of the Industry* for an overview of sheep dairying in Ontario and on E-Mail List to access dairy sheep-oriented Yahoo! groups. Download issues of the ODSA's fine newsletter, *Shepherd's Whey.*

ONTARIO SHEEP MARKETING AGENCY

www.ontariosheep.org
The Ontario Sheep Marketing Agency is a producer-operated organization representing all aspects of the sheep, lamb, and wool industry in Ontario. Its Web site is jam-packed with directories, market and sheep industry news, research reports, consumer tips, lamb recipes, and a *Kid's Corner* (including a cool four-page activity book in PDF format). Click on the *Scrapie Information for Sheep Producers* icon to access late-breaking information about scrapie eradication efforts in North America and abroad.

SHEEP ORGANIZATIONS UNITED KINGDOM
BRITISH COLOURED SHEEP BREEDERS ASSOCIATION

http://www.digitalburn.net/clients/bcsba/
The British Coloured Sheep Breeders Association was formed in 1985 by a group

of UK-based sheep breeders to promote colored sheep and their wool. The Web site offers pictures and world profiles of Britain's 24 breeds of colored sheep (including a dozen available in the United States), along with a wonderful section about fleece—producing it and selling it—and two fantastic charts detailing the fleece qualities of most British-based breeds.

BRITISH SHEEP DAIRYING ASSOCIATION

www.sheepdairying.co.uk
Sheep's milk facts, articles outlining the economics of sheep dairying, recipes, and sheep dairying links are a few of the goodies to be accessed via the British Sheep Dairying Association Web site.

BRITISH WOOL MARKETING BOARD

www.britishwool.org.uk
The British Wool Marketing Board Web site's educational resources (accessible by clicking on fact sheets in the home page menu) are some of the best on the Web. Don't miss *British Sheep Breeds and Their Wool* (a full page of images and information about each of Britain's 60 breeds) and *The Shepherd's Calendar*. Wonderful sheep breed posters, books, even hand-spinners' fleeces can be ordered through the organization's online store.

SHEEP ORGANIZATIONS NEW ZEALAND
NEW ZEALAND SHEEP BREED ASSOCIATION

www.nzsheep.co.nz/index.html
Eighteen breeds raised in New Zealand (and most of them here as well) are pictured and described in detail on the New Zealand Sheep Breed Association Web site. It's a best-bet sheep breeds resource.

RARE BREEDS CONSERVANCIES
AMERICAN LIVESTOCK BREEDS CONSERVANCY/ALBC

www.albc-usa.org
The American Livestock Breeds Conservancy is an organization working to protect nearly 100 breeds of cattle, goats, horses, asses, sheep, swine, and poultry from extinction—23 sheep breeds among them. Visit the site to learn how you can help.

NEW ENGLAND HERITAGE BREEDS CONSERVANCY/NEHBC

www.nehbc.org
The Heritage Breeds Conservancy works to preserve historic and endangered breeds of livestock and poultry. Conservators from across the United States and Canada are listed in the *NEHBC Directory* and participate in the online forum and marketplace listings. A nice selection of conservation-related links rounds out this informative Web site.

RARE BREEDS SURVIVAL TRUST/ RBST

www.rbst.org.uk
The Rare Breeds Survival Trust— the United Kingdom's equivalent of the American Livestock Breeds Conservancy—currently monitors 73 breeds of rare farm animals. Visit the site

to learn more about 29 rare sheep breeds on RBST's Watchlist, including some being concurrently tracked in North America by the American Livestock Breeds Conservancy.

VETERINARY MEDICAL ORGANIZATIONS
AMERICAN ASSOCIATION OF SMALL RUMINANT PRACTITIONERS/ AASRP
www.aasrp.org
The American Association of Small Ruminant Practitioners is a professional organization for veterinarians and veterinary students interested in small ruminant medicine. If you are a veterinarian or you'd like your veterinarian to become involved with sheep on a greater scale, this is your site.

AMERICAN VETERINARY MEDICAL ASSOCIATION/AVMA
www.avma.org
Visit the American Veterinary Medical Association for a peek at the world of veterinary medicine and to view or download articles about hot topics written by veterinarians for the lay public. Scroll to the bottom of the page and click on *NetVet and Electronic Zoo* to access thousands of additional veterinary topic resources.

AMERICAN HOLISTIC VETERINARY MEDICAL ASSOCIATION/ AHVMA
www.ahvma.org
If you're among the growing legion of pet and livestock owners who prefer holistic treatment for your animals, shop the *American Holistic Veterinary Medical Association Referral Directory* to find licensed holistic veterinarians in your locale. The *AHVMA Bookstore* carries hard-to-find print resources. A click on the *Links and Resources* icon leads you to the Web sites of specialty veterinary medical organizations such as the Academy of Veterinary Homeopathy/AVH, the American Veterinary Chiropractic Association, and the Veterinary Botanical Medicine Association.

PHARMACEUTICALS, EQUIPMENT, AND OTHER SHEEP SUPPLIES
AMERICAN LIVESTOCK SUPPLY/ALS
www.americanlivestock.com
(800) 356-0700
American Livestock Supply stocks a full line of sheep (cattle, poultry, swine, etc.) vaccines and equipment at discount prices. Order online or request a print catalog—it's free.

JEFFERS LIVESTOCK SUPPLY
www.jefferslivestock.com/ssc
(800) 533-3377
Jeffers offers the same wide selection of livestock equipment and pharmaceuticals as American Livestock Supply and at competitive prices. Order the free Jeffers catalog—you won't be disappointed.

PIPESTONE VET SUPPLY
www.pipevet.com
(507) 825-5687
Pipestone Veterinary Supply is a division of the Pipestone Veterinary Clinic, where the friendly doctors really know their sheep. Shop the online catalog or request

a free print copy. Pipestone Vet Supply stocks all the standards plus items you won't find anyplace else. While visiting the Pipestone Vet Supply Web site, click on *Archive of Articles* and *Management Tips* to access hundreds of interesting staff-written articles and tips.

PREMIER 1 SUPPLIES

www.premier1supplies.com
(800) 282-6631
Premier has been providing shepherds with fencing, sheep supplies, clippers and shearers, ear tags, and expert advice for more than 25 years. If it's sheepy, Premier sells it—including their own line of clippers and shearers and unique items such as the Premier sheep chair. Shop their comprehensive online store or request Premier's free, tip-filled sheep supply, fencing, and clipper and shearing machine catalogs—all are free.

SHEEPMAN SUPPLY

www.sheepman.com
(800) 331-9122
info@sheepman.com
Sheepman Supply has furnished a full line of shepherds' needs to the industry since 1937. Don't miss their fine selection of sheepy gifts! Order online or request a free print catalog.

SULLIVAN SUPPLY

www.sullivansupply.com
Texas warehouse (800) 588-7096
Iowa warehouse (800) 475-5902
If you show your sheep (or simply want to spiff them up for a special occasion), you need the free Sullivan Supply livestock show supply catalog. Grooming tools and gadgets, blankets, stands, shampoos, and conditioners are but the tip of the Sullivan Supply iceberg.

VALLEY VET SUPPLY

www.valleyvet.com
(800) 419-9524
Like Jeffers and ALS, Valley Vet Supply markets a huge selection of livestock supplies, equipment, and pharmaceuticals at discount prices. Visit the online store or ask for a catalog. Like the others, it's free.

SHEEP COATS

ABC WOOLCRAFTS

www.abcwoolcrafts.mralter.com
The hardest part about coating sheep is keeping their covers patched. ABC Woolcraft's easy-to-use sheep coat repair kit saves shepherds time and hassle and makes sheep covers last longer, too. While you're visiting this site, scope out their nutrition-packed sheep treats: Baa Bars, Ewe Phoria (for pregnant ewes), and Ramma Lamma Ding Dongs. Abram gives the Baa Bars two thumbs up!

MATILDA BRAND LIVESTOCK COVERS FOR SHEEP

www.sheepcovers.com
If you're wondering whether covering your sheep is worthwhile, what to look for in a sheep cover, or how to fit one to your sheep, then the articles at the Matilda Brand Livestock Covers site were written just for you. Matilda Brand sheep covers are manufactured in Australia and recognized as some of the best in the world;

click on *USA Customers* on the home page to find links to North American Matilda Brand dealers.

ROCKY SHEEP COMPANY

www.rockysheep.com
If you'd rather buy American, Rocky Mountain Sheep Company Sheep Suits are made in the United States. Visit the site and read their FAQs to learn more about jacketing sheep.

Fiber Arts

EARTHSONG FIBERS

www.earthsongfibers.com
If a product has anything at all to do with spinning, weaving, dying, knitting, or any of the other fiber arts, Earthsong Fibers probably carries it—and their Web site will demonstrate how to use it.

EBSQ ZINE; SELF-REPRESENTING ARTISTS

www.ebsqart.com
To access superlative step-by-step, photo-illustrated felt-making instructions, visit the EBSQ Zine Web site. Click on *Learn*, then *Live Archives: Felt Making.* They'll have you crafting felt in no time!

THE FELT LADY

www.yurtboutique.com
Suzanne Pufpaff, the Yurt Lady, conducts felting workshops, operates a custom carding mill, and sells felting and knitting supplies. The bibliographies and book reviews on her Web site are invaluable tools for the beginning felter. Her felting links page is especially useful.

FELTMAKERS LIST FAQ

www.peak.org/~spark//feltlistFAQ.html
If you want to learn feltcrafting, you mustn't miss this page. Subscribe to the Feltmakers e-mail discussion list, peruse the list's comprehensive FAQs, or follow dozens of how-to resource links to the best felting material on the Internet.

HANDWEAVER'S GUILD OF AMERICA

www.weavespindye.org
Founded in 1969 to inspire creativity and encourage excellence in the fiber arts, the Handweavers Guild of America unites weavers, spinners, dyers, basket makers, fiber artists, and educators. Join the organization or locate a local guild via their Web site.

INTERWEAVE PRESS

www.interweave.com
Interweave Press publishes books and magazines devoted to spinning, knitting, weaving, fiber arts, needlework, beading, and natural living. Visit the Web site to view their catalog, subscribe to a magazine, or locate a nearby bookseller selling Interweave titles. Click on a topic for a marvelous review of each, including a host of links to online resources; the *Spinning* category is especially resource rich. This site will keep you happily occupied for days!

JOY OF HANDSPINNING

www.joyofhandspinning.com
Joy of Handspinning lives up to its billing as "The Web site for handspinners." View streaming and audio demonstrations, read hundreds of how-to articles, pur-

chase hard-to-find books, and subscribe to Joy of Handspinning's free e-mail list at this absolutely must-visit Web site.

KINGDOM OF EALDORMERE, SPINDLE SPINNING

www.ealdormere.sca.org/university/spin-dlespinning.shtml

Everything you need to know to start drop-spindle spinning (and then some) is on this fascinating Web page.

LAST OF THE FELTMAKERS

www.atamanhotel.com/felt.html

See how felt is made in the traditional manner at this interesting Turkish site.

URBAN SPINNER

www.urbanspinner.com

Elaine Benfatto hosts this gorgeous, noncommercial guide to handspinning resources on the World Wide Web. Be sure to click on *Handspinning Links* to access more handspinning Web sites than you probably imagined existed.

PET SHEEP TRAINING AND ANIMAL THERAPY
KAREN PRYOR'S CLICKER TRAINING

www.clickertraining.com

Karen Pryor wrote the book (*Don't Shoot the Dog*) that brought clicker training to the masses. She is still at the forefront of the clicker training movement. If you don't know what clicker training is, this is the place to learn. Sheep are easily trained using methods tailored for horses and dogs. The information on this Web site is enough to get you started training sheep.

CLICKER SOLUTIONS

www.clickersolutions.com

Clicker Solutions is both a Web site and an e-mail list. Visit the Web site to subscribe to the list (it's free) or peruse the many clicker-training articles and FAQs you'll find linked to this page.

PET PARTNERS PROGRAM

www.petpartners.org

If you'd like to do animal-assisted therapy with your pet sheep, the Pet Partners Program is for you. Visit the Web site for particulars and to enroll.

FLOCK GUARDIANS
LIVESTOCK GUARDIAN DOGS (LGDS)

www.lgd.org

Everything about Livestock Guardian Dogs is here. Subscribe to the LGD-Lovers (LGD-L) E-mail list, browse the list's FAQs, and click on *Library* to go to hundreds of informative articles about dogs in general and LGDs in particular. This site is brought to you by the Livestock Guardian Dog Association—don't miss it!

FLOCK AND FAMILY GUARDIAN DOG COMPREHENSIVE RESOURCE GATEWAY

www.flockguard.org

A second must-visit site is the Flock and Family Guardian Dog Comprehensive Resource Gateway. Access e-mail lists and free newsletters, as well as guardian breed rescue sites, and view photo galleries and interesting articles about guardian dog breeds. Lots of links to other guardian dog sites are icing on the cake.

PUT A LLAMA IN YOUR LIFE

www.llama.org

Surf this comprehensive site to learn all about llamas. Scroll down the home page to *What Do You Do With a Llama* and click on *Guarding Livestock* for the skinny on llamas as flock guardians.

USING DONKEYS TO GUARD SHEEP AND GOATS

www.agr.state.tx.us/pesticide/brochures/pes_donkeys.htm

Brought to you by the Texas Department of Agriculture, this bulletin is the Web's best guide to choosing and using donkeys as flock guardians.

LAMB AND MUTTON RECIPES
AMERICAN LAMB BOARD'S LAMB CHEF

www.lambchef.com

The American Lamb Board brings you an enormous searchable database of lamb recipes and two recipe brochures in downloadable PDF format.

LAMB RECIPES AT JUST SLOW COOKING

www.justslowcooking.com/inxlam.html

Just Slow Cooking brings you 58 yummy lamb and mutton recipes you can fix in a slow cooker, more than any other recipe site on the Web.

LAMB RECIPES AT RECIPE HOUND

www.recipehound.com/Recipes/lamb.html

At Recipe Hound, you'll find more than 100 lamb recipes ranging from simple (Lamb Stew) to sublime (Basque Leg of Lamb with Mushroom and Wine Sauce) to somewhat strange (Lamb Tongues in Madeira Sauce).

MCKEAN FAMILY OF HAGGIS HOME PAGE

www.scottishhaggis.co.uk

If you love haggis, order it directly from Scotland's McKean Family of Haggis, master haggis makers since 1850. A click on *Cooking a Haggis* in the left-hand menu shows in word and image how to prepare this Scottish delicacy. If you'd rather fling than eat it, click on *Haggis Hurling* instead.

SHEEP'S CREEK FARM LAMB RECIPES

www.sheepscreek.com/recipe_list.html

Sheep's Creek Farm brings you a comprehensive list of more than 100 recipes featuring lamb and mutton.

University Resources

Major state universities and state extension services distribute papers and bulletins of interest to shepherds. While visiting these Web sites, check for other useful bulletins under headings such as Farm Construction, Forage (hay and pasture), and Poisonous Plants. To compile an up-to-date free library of sheep materials, download appropriate PDF files to save for future reference, print favorite bulletins and file them, or bind printouts to create your own personal "everything about sheep" reference book.

AGMRC/AGRICULTURAL MARKETING RESOURCE CENTER

www.agmrc.org

AgMRC comprises marketing experts from Iowa State University, Kansas State University, and the University of California working together to create and disseminate information about value-added agriculture. Small-scale sheep and wool producers: don't miss this valuable site! Visit AgMRC to explore marketing trends, e-mail your questions to AgMRC specialists, and peruse thousands of valuable print and online resources. Access lamb resources by clicking on *Commodities & Products*, then *Livestock*, then *Lamb*.

ALABAMA COOPERATIVE EXTENSION SYSTEM

www.aces.edu

The Alabama Cooperative System is an especially rich source of sheep management resources. To access them, click on Livestock, then *Animal Science* Extension Programs, then *Sheep* and *Goat Production*. Hundreds of bulletins are listed.

UNIVERSITY OF ARIZONA COLLEGE OF AGRICULTURAL LIFE AND SCIENCES

www.cals.arizona.edu

To navigate the University of Arizona College of Agricultural Life and Sciences site, click on *Extension/Outreach*, then *Animals,* then *Sheep*.

CLEMSON UNIVERSITY EXTENSION

hgic.clemson.edu

Clemson University offers great sheep bulletins in PDF format. Click on *Extension Home*, then *Publications* (at the top of the page), then *Digital Publications Only*. Next, click on *Animal and Veterinary* Sciences for general bulletins or 4-H to download an excellent 4-H show-lamb guide.

COLORADO STATE UNIVERSITY COOPERATIVE RESOURCE CENTER

www.cerc.colostate.edu

The Colorado State University Cooperative Resource Center publishes a huge selection of useful shepherds' bulletins; to access them click on *Titles* and then *Livestock Publications*.

CORNELL SHEEP PROGRAM

www.sheep.cornell.edu

The Cornell Sheep Program Web site is one of the best sources of university-generated shepherds' resources on the Internet. Don't miss the links pages; they are extensive.

UNIVERSITY OF GEORGIA COLLEGE OF AGRICULTURAL AND ENVIRONMENTAL SCIENCES COOPERATIVE EXTENSION SERVICE

extension.caes.uga.edu

The University of Georgia College of Agricultural and Environmental Sciences publishes thousands of online and PDF-format documents. To locate sheep and agriculture-related titles, click on *Publications*, then *Subject Listing*, and scroll down to your topics of interest.

UNIVERSITY OF ILLINOIS AT URBANA-CHAMPAIGN; TECHNOLOGY AND RESEARCH: ALLIED AND INTEGRATED FOR LIVESTOCK LINKAGES

Access a plethora of university-generated sheep bulletins by clicking on *SheepNet*, then *Resources*. Click on Historical at the SheepNet page for an interesting peek into sheep-keeping practices of the past.

IOWA STATE UNIVERSITY EXTENSION

www.extension.iastate.edu

Iowa State University Extension offers shepherds a dozen exceptionally well-written sheep bulletins in PDF format. To find them, click on *Publications*, then *Livestock*, then scroll down to the sheep resource listings. You'll find a world of other useful bulletins on this Web site, so take your time and shop around when you visit.

K STATE RESEARCH AND EXTENSION KANSAS STATE CENTER FOR SUSTAINABLE AGRICULTURE AND ALTERNATIVE CROPS

www.oznet.ksu.edu

You'll find not only Kansas State University's excellent sheep bulletins but the best sheep bulletins published by several other university extensions on this easily navigated site. Click on *Publications*, then *S*, then *Sheep*.

UNIVERSITY OF KENTUCKY COLLEGE OF AGRICULTURE

www.ca.uky.edu/agripedia

You'll find two fine sheep features on the University of Kentucky College of Agriculture's cool Agripedia site. For a quick, fun peek at major sheep breeds, click on the *Agrimania* pull-down menu and select *Livestock Breeds*, click *Sheep*, then *Meat Breeds*, *Wool Breeds*, or *Dual Purpose Breeds*. To access an array of useful university bulletins, instead click *Subject Index*, then click *S* (for sheep), or simply scroll through the vast list of available agricultural publications.

UNIVERSITY OF MAINE COOPERATIVE EXTENSION

www.umext.maine.edu

The University of Maine Cooperative Extension offers several useful sheep bulletins downloadable in PDF format. To reach them click on *Publications*, then *Online Catalog*, then *Agriculture—Dairy, Livestock and Poultry*.

UNIVERSITY OF MARYLAND'S WESTERN MARYLAND RESEARCH AND EDUCATION CENTER

www.westernmaryland.umd.edu

Accessing the huge selection of sheep resources available through the University of Maryland's Western Maryland Research and Education Center is the essence of simplicity: click on Sheep & Goats in the menu, and there you are. Peruse or download *Maryland Sheep and Goat Producer* newsletters, and access the Maryland Small

Ruminants Page, the Northeast Sheep and Goat Marketing Web site, and a lot of other useful sheep-oriented materials via this Web site.

MICHIGAN STATE UNIVERSITY EXTENSION

cvm.msu.edu

To access Michigan State University Extension's excellent array of sheep bulletins and links to other universities' sheep bulletins, click on *Information Resources,* then *Information Access Center,* then *Animals,* then *Sheep.*

UNIVERSITY OF MINNESOTA DEPARTMENT OF ANIMAL SCIENCES

www.ansci.umn.edu

University of Minnesota Department of Animal Sciences Web site is a must-visit due to its ease of navigation and its rich storehouse of sheep-specific resources. A click on *Sheep* in the left-hand menu takes you to links to University of Minnesota sheep bulletins and a fully downloadable version of their 83-page *Sheep Care and Management Manual.*

UNIVERSITY OF MINNESOTA EXTENSION SERVICE

www.extension.umn.edu

Some, but not all, of the above publications are also available from the University of Minnesota Extension Service Web site, along with more than two dozen other interesting sheep-related bulletins. Access them by clicking Farm, then *Sheep* in the left-hand menus.

UNIVERSITY OF MISSOURI EXTENSION

muextension.missouri.edu

University of Missouri Extension offers a selection of useful sheep bulletins. Access them by clicking on *Publications*, then *Agriculture*, then *Sheep.*

MONTANA STATE UNIVERSITY ANIMAL AND RANGE SCIENCE EXTENSION

www.animalrangeextension.montana.edu

The Montana State University Animal and Range Science Extension Web site is home to market reports, the entire *Montana Farm Flock Sheep Production Handbook*, and a fine collection of sheep articles. Click on *Sheep* to access the main portal, then *FactSheets.*

UNIVERSITY OF NEBRASKA-LINCOLN INSTITUTE OF AGRICULTURE AND NATURAL RESOURCES

ianrpubs.unl.edu

Don't miss the collection of great sheep articles at the University of Nebraska-Lincoln Institute of Agriculture and Natural Resources Web site. Reach them by clicking on *Sheep* under *Browse Publications.*

NORTH CAROLINA STATE UNIVERSITY COLLEGE OF AGRICULTURE LIFE AND SCIENCE

www.cals.ncsu.edu

Navigate to the North Carolina State University College of Agriculture Life and Science's sheep bulletins and resources by clicking on *Extension*, then *Agriculture & Food*, then *Animal Agriculture*, then *Animal Husbandry*, and finally *Sheep*. Also click on *4-H Youth Livestock* at the *Animal Husbandry* page, then *Sheep* to access several great articles about showing club lambs; don't miss the one on Market Lamb Showmanship.

NORTH DAKOTA STATE UNIVERSITY EXTENSION SERVICE

www.ext.nodak.edu

There's lots to like at the North Dakota State University Extension Service Web site. Click on *Livestock*, then *Sheep Publications*—navigation is that easy. Also click on N*orth Dakota Lamb and Wool Industry Newsletter* to view loads of archived newsletters.

OKLAHOMA STATE UNIVERSITY AGRICULTURALCOMMUNICATIONS SERVICES

osuextra.okstate.edu

Access Oklahoma State University's excellent bulletins, all of them downloadable in PDF format, by clicking on *Animals*, then *Sheep*.

OREGON STATE UNIVERSITY EXTENSION AND EXPERIMENT STATION

eesc.orst.edu

Oregon State University Extension and Experiment Station offers only a few sheep bulletins, but they are good ones. To access them, click on *Publications & Videos*, then *Agriculture*, and finally *Horses, Sheep, Goats and Swine*.

ANIMAL EXTENSION SERVICES @ PURDUE UNIVERSITY

ag.ansc.purdue.edu/anscext

You'll find some of the Internet's most comprehensive sheep resources pages at the Animal Extension Services @ Purdue University Web site. Click on *Species Information,* then *Sheep*, and then *Articles, Student Developed Web Fact Sheets,* or *Sheep Links*.

SOUTH DAKOTA STATE UNIVERSITY COLLEGE OF AGRICULTURE AND BIOLOGICAL SCIENCES COOPERATIVE EXTENSION SERVICE

sdces.sdstate.edu

To access South Dakota State University College of Agriculture and Biological Sciences Cooperative Extension Service sheep bulletins, select *Sheep* in the pulldown menu under *Livestock*—and there you are!

UNIVERSITY OF TENNESSEE EXTENSION

www.utextension.utk.edu

To navigate your way to the fine sheep bulletins offered by the University of Tennessee Extension, click on *Animals & Livestock*, then *Sheep Programs*. Don't miss the invaluable 24-page PDF download *Applied Sheep Behavior*—it's a honey!

TEXAS A&M UNIVERSITY DEPARTMENT OF ANIMAL SCIENCE

animalscience.tamu.edu

Texas A&M University Department of Animal Science online sheep resources are easy to access: click on *Sheep* and *Goats*, then the *Sheep and Goats Publications and Information Center* icon. All Texas A&M sheep bulletins are formatted as handy PDF downloads.

UTAH STATE UNIVERSITY EXTENSION

www.extension.usu.edu

The many sheep bulletins available through the Utah State University Extension are harder to access than most, but they're good ones and worth the effort. To find them, click on *Publications*, then enter one-word searches for *sheep, rams, ewes*, and *lambs*.

VIRGINIA COOPERATIVE EXTENSION

www.ext.vt.edu

Virginia Cooperative Extension sheep bulletins are among the best university-generated sheep resources on the Web. To reach them, click on *Educational Programs & Resources*, then *Livestock, Poultry & Dairy*, and finally *Sheep*.

UNIVERSITY OF WISCONSIN SHEEP EXTENSION

www.uwex.edu/ces/animalscience/sheep

Sheep are big business in Wisconsin— and this five-star Web site is one of the most comprehensive university sheep sites on the Internet. Visit to learn about events such as Spooner Sheep Day, the Great Lakes Dairy Sheep Symposium, and UWEX's shearing school. Then click on *Publications & Proceedings* to access hundreds of sheep-related university bulletins.

Government Resources

UNITED STATES

NATIONAL INSTITUTE OF ANIMAL AGRICULTURE

www.animalagriculture.org

The National Institute of Animal Agriculture provides individuals and organizations with information, education and solutions for the challenges facing animal agriculture. To download archived issues of the *Sheep and Goat Health Report* and to access the organization's extensive scrapie resources, simply click on the smiling sheep icon.

USDA ANIMAL AND PLANT HEALTH INSPECTION SERVICE

www.aphis.usda.gov

Anyone who keeps sheep must enroll his or her flock in one of two federally mandated scrapie-eradication programs. To peruse the latest official scrapie news and program specifics, click on *Programs*, then N*ASFA Safeguarding Report* (under Veterinary Services), then *Sheep* in the left-hand menu.

USDA AGRICULTURAL MARKETING SERVICE/
USDA NATIONAL ORGANIC PROGRAM

www.ams.usda.gov

If you'd like to market organic lamb and are seeking the official word on organic certification, visit the Agricultural Marketing Service Web site. Click on *National Organic Program*, then your topic of interest.

AUSTRALIA
NEW SOUTH WALES DEPARTMENT OF PRIMARY INDUSTRIES
www.agric.new.gov.au/reader
To access the excellent sheep resources offered by the New South Wales Department of Primary Industries Web site, place your curser over *Livestock* in the left-hand menu, then click on *Sheep*.

QUEENSLAND GOVERNMENT/ DEPARTMENT OF PRIMARY INDUSTRIES AND FISHERIES
www.dpi.qld.gove.au/home/default.html
Navigate to the Queensland government's Department of Primary Industries and Fisheries sheep resources page by clicking on *Sheep* in the left-hand menu.

CANADA
BRITISH COLUMBIA MINISTRY OF AGRICULTURE, FOOD AND FISHERIES
www.agf.gov.bc.ca
The government of British Columbia's large collection of excellent sheep bulletins can be accessed by clicking on *Reports & Publications,* then *Publications Available on the BCMAFF Website—Full Listing,* and finally, *Sheep.*

CANADA PLAN SERVICE
www.cps.gov.on.ca
Canada Plan Service is a network of agricultural engineers and livestock specialists involved in gathering ideas from across Canada, then developing construction and management recommendations for new farm structures based on their findings. Although detailed plans can be ordered at a nominal cost, PDF files of plans for hundreds of buildings and items such as livestock-handling equipment and feeders can be downloaded from the site free of charge.

ONTARIO MINISTRY OF AGRICULTURE AND FOOD
www.gov.on.ca/OMAFRA
The Ontario Ministry of Agriculture and Food Web site is a rich source of information for shepherds large and small. To access hundreds of management titles, click on *Agriculture,* then *Livestock*, then scroll down the page to *Sheep.*

SASKATCHEWAN AGRICULTURE, FOOD AND RURAL REVITALIZATION
www.agr.gov.sk.ca
The excellent sheep documents archived at the Saskatchewan Agriculture, Food and Rural Revitalization Web site can be accessed by clicking on *Agriculture,* then *Livestock*, then *Sheep and Goats.*

Other Useful Web Sites
ALL ABOUT SHEEP FOR KIDS
www.kiddyhouse.com/farm/sheep
If there are children in your family, they'll love this noncommercial, kid-oriented guide to sheep, where they'll learn about the history of sheep, hear sheep *baa*, view and read about common sheep breeds, print out sheep coloring pages, and follow links to kids' sheep-themed songs, clip art, craft activities, story time, and more.

AMERICAN GRASSFED ASSOCIATION

www.americangrassfed.org

Sheep producers interested in niche marketing grass-fed lamb will find much to think about on this well-organized site.

EAT WILD—THE CLEARINGHOUSE FOR INFORMATION ABOUT PASTURE-BASED FARMING

www.eatwild.com

The goal of this Web site is to "provide comprehensive, up-to-date, information about the benefits of choosing beef, pork, lamb, bison, poultry, and dairy products from pastured animals." If you're considering raising grass-fed lamb, visit this valuable resource to learn the why and how of pasture-based farming.

FOOD AND AGRICULTURE ORGANIZATION OF THE UNITED NATIONS/FAO

www.fao.org/docrep/v9384e/v9384e00.htm

Since its founding in 1945, the Food and Agriculture Organization of the United Nations has focused special attention on developing rural areas, home to 70 percent of the world's poor and hungry people. To that end, the FAO has produced thousands of excellent, well-illustrated how-to manuals, including *Harvesting of Textile Animal Fibres.* If you'd like to shear your own sheep, don't miss this valuable online resource.

HOBBY FARMS SHEEP

www.geocities.com/HobbyFarmsSheep

Visit the author's Hobby Farms Sheep Web site, "An online resource guide for hobby farm shepherds."

JOHNE'S INFORMATION CENTER

www.johnes.org

The University of Wisconsin School of Veterinary Medicine bills this site as "Your definitive source for information on Johne's Disease." To access sheep-specific resources, use the pull-down menu to specify *Sheep,* then click on the topic you'd like to peruse.

KERR CENTER FOR SUSTAINABLE AGRICULTURE

www.kerrcenter.com

The Kerr Center publishes educational materials on a wide range of sustainable farming and ranching, alternative marketing, food and agriculture policy, and rural development topics. Of special interest to shepherds is *More Profit With Hair Sheep* by Dr. Gerald Fitch. To download it as a PDF file, click on *Publications,* then *Farming, Ranching, Marketing, Alternative Income,* and scroll on down to *Goats, Pigs, Poultry, Sheep.* Surf the site while you're there; you'll find loads more bulletins of interest to hobby farmers.

MARYLAND SMALL RUMINANT PAGE

www.sheepandgoat.com

Susan Schoenian, Sheep and Goat Specialist for the University of Maryland Cooperative Extension, hosts this amazing collection of original documents and links to thousands of additional online resources. Virtually anything you want to know about sheep can be accessed from this site.

NATIONAL SUSTAINABLE AGRICULTURE INFORMATION SERVICE/ATTRA

attra.ncat.org

ATTRA—National Sustainable Agriculture Information Service, funded by the US Department of Agriculture, is managed by the National Center for Appropriate Technology. It provides information and other technical assistance to farmers, ranchers, extension agents, educators, and others involved in sustainable agriculture throughout the United States. To access ATTRA's sustainable sheep production and sheep dairying bulletins (read them online or download them as PDF files), click on *Livestock*, then *Hogs, Sheep, and Goats*. You'll find oodles of marketing and record-keeping resources here too.

OPP CONCERNED SHEEP BREEDERS SOCIETY

www.oppsociety.org

For the lowdown on Ovine Progressive Pneumonia, use the OPP Concerned Sheep Breeders Society's pull-down menu to navigate to your OPP topic of choice.

SHEEP 101

www.sheep101.info

Also maintained by Susan Schoenian of the Maryland Ruminant Page, this great site answers common sheep questions in words and beautiful images. An exceptionally fine resource for older children.

Print Resources

The following fine books and periodicals are written for small-scale hobby-farm shepherds rather than big biz meat and wool producers.

Most sheep supply retailers and fiber art supply catalogs carry sheep books, and you'll find in-print titles at Amazon (www.amazon.com) and Barnes and Noble (www.barnesandnoble.com), but eBay (www.eBay.com) remains a best source for sheep books, especially for collectors of out-of-print, British, and scarce titles.

SHEEP BOOKS

The Beginning Shepherd's Manual
Barbara Smith, Mark Aseltine, and Gerald Kenney, 2d ed. (Blackwell Publishing, 1997)
The title says it all. The book is written by an experienced sheep producer, with chapters by a ruminant nutritionist and a veterinarian.

British Sheep Breeds
Elizabeth Henson (Shire Publications, 2000)
Many of our sheep breeds originally hailed from Great Britain or were developed using British genetics. This slim but information-packed volume in the Shire Album series describes 48 British breeds in text and excellent black-and-white pictures.

British Sheep and Wool
Edited by J. Elliot, D. E. Lord, and J. M. Williams, rev. (British Wool Marketing Board, 1990)

This 112-page, full-color oversize paperback is our favorite wish book. This edition pictures and describes 120 breeds of sheep and discusses the types of woolen products woven from their fleeces. The book is divided into sections: Main Breeds, Minor Breeds, Rare Breeds, Hybrids and Halfbreds, Recent Introductions, and Southern Hemisphere Sheep. Main breeds each garner a two-page spread: a full page flock view facing another full page including a breed description and a large picture of an individual sheep. The photography is glorious, the descriptions fascinating; it's inexpensive and available in North America through Premier's online and print sheep supply catalogs.

Handspun Treasures from Rare Wools: Collected Works from the Save the Sheep Project
Edited by Deborah Robson (Interweave Press; 2000)
If you market handspinners' fleeces, you'll love this book! Its ninety-six pages show (in black-and-white and gorgeous color) dozens of handcrafted pieces made of wool, along with pictures and descriptions of the rare breed sheep from which it came. It discusses rare breed conservation and the history of handspinning, and it teaches several basic handspinning techniques. The resources section is outstanding.

Healthy Sheep Naturally
Pat Coleby, rev. (Landlinks Press, 2000)
This hard-to-find 184-page Australian paperback is one of few references detailing the holistic production of sheep. The herbal, homeopathic, and natural remedies chapter is worth the price of the book.

Homeopathy: The Shepherd's Guide
Mark Elliot and Tony Pinkus (1993)
If you're interested in homeopathics, you need this fact-packed 32-page British paperback. Buy it new in North America from Whole Health Now (http://www.wholehealthnow.com), 1642 Fickle Hill Road, Arcata, CA 95521, (707) 822-5807—or watch for it used on eBay.

In Sheep's Clothing: A Handspinner's Guide to Wool
Nola and Jane Fournier (Interweave Press, 2001)
This book is a bonanza for handspinners and shepherds who sell handspinners' fleeces. The book describes (in detail) the fleece qualities of 94 breeds of sheep—fiber diameter in microns, spinning counts, staple length, fleece weight, and more—and pictures a full-size lock of typical fleece. The book also offers advice on selecting top-quality fleeces; cleaning wool efficiently and thoroughly; teasing, flicking, combing, carding, and other preparation methods; and spinning and plying a variety of yarn styles.

Lamb Problems: Detecting, Diagnosing, Treating
Laura Lawson, rev. (LDF Publications, 1996)
Managing Your Ewe and Her Newborn Lambs
Laura Lawson, rev. (LDF Publications, 1997)
Laura Lawson authors and self-pub-

lishes several of the most valuable sheep books on the market. Using diagnostic check sheets, symptom flowcharts, text, and illustrations, she guides the reader through the nuances of diagnosing and treating most any problem associated with breeding and caring for sheep and lambs. These are thick, well-written volumes; they deserve a place of honor in every lambing kit and on every shepherd's bookshelf. Available from Laura Lawson (http://www.tias.com/stores/sbdc), 11114 Lawson Lane, Culpeper, VA 22701.

Practical Sheep Keeping
Kim Cardell (Crowood Press, 1998)
This is our favorite of many excellent small-scale sheep-keeping manuals published in Great Britain. Its 160 pages are jam-packed with information new shepherds need to know, from selecting a breed to lambing to marketing. It's hard to locate in North America but well worth the search—ours came to us secondhand via eBay.

Raising Sheep the Modern Way
Paula Simmons (Garden Way Publishing, 1998)
This out-of-print classic has been updated and reprinted as *Storey's Guide to Raising Sheep*. However, the original *Raising Sheep the Modern Way* is a complete introduction to sheep raising and a still-useful addition to the shepherd's bookshelf.

The Shepherd's Guidebook
Margaret Bradbury (Rodale Press; 1977)
Although a small proportion of the information in this out-of-print book is dated, it's still a new shepherd's best buy. This 222-page volume covers everything from starting a flock to breeds and breeding, from pet lambs to home butchering (with recipes such as Swedish Lamb Shanks, Lancashire Hot Pot, and Lamb Loaf). The housing and sheep equipment diagrams are especially worthwhile.

Sheep (Complete Pet Owner's Manual)
Hans Alfred Muller (Barron's, 1989)
This is a seventy-two-page quality paperback originally published in German as *Shafe als Haustiere*. Its size is deceiving—this slim volume is jam-packed with useful information for pet sheep and small flock owners. Its "The Behavior of Sheep" chapter is outstanding, as are the gorgeous color pictures and line drawings that liberally pepper this work. It's one of our favorite sheep books.

The Sheep Book; A Handbook for the Modern Shepherd
Ron Parker, rev. (Ohio University Press, 2001)
The *Sheep Book* is a completely updated 321-page, profusely illustrated revision of Ron Parker's 1983 classic of the same name. Because they cover most every conceivable sheepy topic, it and *Storey's Guide to Raising Sheep: Breeds, Care, Facilities*, are our sheep-keeping standbys. Out-of-print copies of the 1983 edition are frequently listed at eBay, or—thanks to the author's generosity—shepherds can download the entire book in PDF format at: http://hem.bredband.net/ronpar/tsb.html

Showing Sheep: A Selecting, Raising, Fitting and Showing Guide
Laura Lawson (LDF Publications; 1994)
This information-packed 224-page manual covers buying or raising a show-quality lamb, then feeding, fitting, training and exhibiting it in it the show ring. *Showing Sheep* is an invaluable reference for both youth (4-H and FFA) and adult showers. Buy it direct from the author at the address given under *Lamb Problems: Detecting, Diagnosing, Treating*.

Small Scale Sheep Keeping
Jeremy Hunt (Faber & Faber, 1997)
In Britain, small-scale sheep keeping is the norm. Many British books are best bets for American shepherds seeking reliable information about small flock selection and management. Jeremy Hunt's *Small Scale Sheep Keeping* is an especially good buy.

Storey's Guide to Raising Sheep: Breeds, Care, Facilities
Paula Simmons and Carol Ekarius (Storey Books, 2001)
If you buy only one sheep book (in addition to the one you're reading), choose *Storey's Guide to Raising Sheep*. "Comprehensive" scarcely describes this up-to-date, 390-page volume. If it's about sheep, it's in there, along with clear line drawings and black-and-white photos illustrating every process. A note for dedicated handspinners: you'll find excellent make-your-own sheep cover diagrams.

Turning Wool into a Cottage Industry
Paula Simmons (Storey Publishing, 1991)
This is an invaluable reference for those longing to earn a living with sheep. Chapters include "Wool as a Cottage Industry," "Sheep Breeds and Crossbreeds," "Sheep Management," "Preparing and Selling Raw Wool," and "Preparing and Selling Washed Wool," along with the lowdown on carding as a business, business and merchandising tips, helpful questions and answers, and profiles of successful home wool business entrepreneurs.

TV Vet Sheep Book: Recognition and Treatment of Common Sheep Ailments
The TV Vet [Eddie Straiton], rev. (Farming Press, 1992)
Because it's illustrated with sequential photos, many shepherds consider this 198-page British work the best sheep veterinary volume on the market. It's really good—we just like *The Veterinary Book for Sheep Farmers* a wee bit better.

The Veterinary Book for Sheep Farmers
David C. Henderson (Farming Press, 1997)
This comprehensive British book is our favorite veterinary reference. Henderson writes in easily understandable layman's terms, and covers every conceivable sheep health topic. If you live where sheep-savvy veterinarians are scarce, you *need* this book!

A Veterinary Guide for Animal Owners: Cattle, Goats, Sheep, Horses, Pigs, Poultry, Rabbits, Dogs, Cats
C.E. Spaulding, D.V.M. (St. Martins Press, reissue 1997)
While this book isn't sheep-specific, the

sheep chapter in *A Veterinary Guide for Animal Owners* is a good one. Handling emergencies, diagnosing problems, coping until the vet arrives—Dr. Spaulding walks you through the process in concise, easy-to-understand text and line drawings.

Your Sheep: A Kid's Guide to Raising and Showing
Paula Simmons, Darrel L. Salsbury (Storey Books, 1992)
Your Sheep: A Kid's Guide to Raising and Showing is essentially *Storey's Guide to Raising Sheep* rewritten for kids— but with added topics (such as detailed sheep-showing how-to, sheep equipment building instructions, and wool craft projects) that make this 120-page, seven-by-ten-inch quality paperback a must-have for adult shepherds, too.

RARE BREED BOOKS
The Encyclopedia of Historic and Endangered Livestock and Poultry Breeds
Janet Vorwald Dohner (Yale University Press, 2001)
This reference book is a huge volume that discusses in depth the numerous merits of rare breed conservation and profiles nearly 200 breeds of livestock (goats, sheep, swine, cattle, horses, and asses) and poultry (chickens, turkeys, ducks, and geese).

A Rare Breeds Album of American Livestock
Carolyn J. Christman, et al. (American Livestock Breeds Conservancy; 1998)
Sheep conservators will also want *A Rare Breeds Album of American Livestock*, 118 pages about the American Livestock Breed Conservancy and other preservation groups' efforts to save endangered livestock breeds—including sheep.

PERIODICALS
Most sheep organizations (and universities in sheep production states) publish newsletters of interest to hobby-farm shepherds. Check them out; you can read samples and archived back issues of many such sheep newsletters online.

Acres USA: A Voice for Eco-Agriculture
www.acresusa.com
PO Box 91299
Austin TX 78709-5299
(800) 355-5313
Grass-fed and organic lamb producers will love this monthly tabloid. Visit the Web site and fill out a form, and you'll receive a free issue and the *Acres USA* book and video catalog. While you're there, click on *Feature Article Archives* to view dozens of articles.

Black Sheep Newsletter
members.aol.com/jkbsnweb
25455 NW Dixie Mtn Rd
Scappoose OR 97056
(503) 621-3063
The *Black Sheep Newsletter* is a friendly, reader-written quarterly for sheep growers and fiber enthusiasts. Issues numbers 1 through 22 are collected in a book titled *Black Sheep Newsletter Companion*; it and most other back issues are available from the publisher, as are additional books, T-shirts, and other promotional items.

sheep!

www.sheepmagazine.com
W11564 Hwy 64
Withee WI 54498
(715) 785-7979

Billed as "The Voice of the Independent Flockmaster," *sheep!* is an information-packed, homey bimonthly published by the same company that publishes *Countryside Magazine*. Visit the Web site to view the entire current issue of *sheep!* along with select archived material from issues past. If you're a small-scale shepherd and you subscribe to just one sheep periodical, choose sheep!

Sheep Connection

http://sheepconnection.tripod.com
2145 Megee Lane
Nicholasville KY 40365

Members of Georgia, Kentucky, South Carolina, and Tennessee state sheep and wool producers associations receive *Sheep Connection* as a membership bonus. Shepherds outside these states can subscribe to it via its Web site.

The Shepherd's Journal

http://www.shepherdsjournal.com
Box 383
Delia AB T0J 0W0
Canada

The Shepherd's Journal publishes a cross section of articles for novice- through veteran-level shepherds. Most relate to the Canadian sheep industry but are applicable to American sheep production as well. Visit the Web site to subscribe or to peruse the excellent Breed Profile feature: select breeds from the thirty-plus listed in *The Shepherd's Journal's* comprehensive pull-down menu to view detailed profiles of each.

The Stockman Grass Farmer: "The Grazier's Edge"

stockmangrassfarmer.com/sgf
282 Commerce Park Drive
Ridgeland, MS 39157

Call (800) 748-9808 or fill in the online form to receive a free sample issue of *The Stockman Grass Farmer*. Published twelve times a year, it serves as an information network for grassland farmers. Grass-fed lamb producers won't want to miss the great archived articles and FAQs on this Web site.

Photo Credits

Cover
juliamcc/Shutterstock.com

Front Matter
John and Sue Weaver 2
Norvia Behling/Paulette Johnson 3
Agricultural Research Service, United
States Department of Agriculture 6

Introduction: Why Sheep?
Photodisc, Inc. 8

Ch 1: Sheep from the Beginning
Photodisc, Inc. 10, 14, 17–18
John and Sue Weaver 12–13, 16, 19,
22–23
JupiterImages and its Licensors 20
Agricultural Research Service,
United States Department of
Agriculture 21

Ch 2: Buying the Right Sheep for You
Agricultural Research Service, United
States Department of Agriculture 24
John and Sue Weaver 27, 30–33, 35
JupiterImages and its Licensors 29, 34

**Ch 3: Housing, Feeding, and Guarding
Your New Flock**
Agricultural Research Service, United
States Department of Agriculture 36
JupiterImages and its Licensors 38, 44
John and Sue Weaver 39–41, 45–47
Photodisc, Inc. 43

**Ch 4: Sheepish Behavior and Safe
Handling**
John and Sue Weaver 48, 52, 54, 56
JupiterImages and its Licensors 50–51
Agricultural Research Service, United
States Department of Agriculture 53

Index

ABOUT THE AUTHOR

Sue Weaver has written hundreds of articles about animals over the years, is a contributing editor of *Hobby Farms* magazine, and is the author of *Chickens: Tending a Small-Scale Flock*. Sue maintains a flock of chickens that includes Barred Rock Cochins, a Silver-Laced Wyandotte, a larger brown Cochin, a Red Jungle Fowl, and an assortment of barnies. She also breeds Keyrrey–Shee miniature sheep, American Curly horses, and AMHR miniature horses of the cob type. Sue lives in Mammoth Spring, Arkansas.